An Introduction
to Regression Graphics

An Introduction to Regression Graphics

R. Dennis Cook
Sanford Weisberg
The University of Minnesota
St. Paul, Minnesota

A Wiley-Interscience Publication

JOHN WILEY & SONS, INC.

New York   Chichester   Brisbane   Toronto   Singapore

Library of Congress Cataloging in Publication Data

Cook, R. Dennis.
    An introduction to regression graphics / R. Dennis Cook, Sanford
  Weisberg.
        p. cm.— (Wiley series in probability and mathematical
  statistics. Probability and mathematical statistics)
    "A Wiley-Interscience publication."
    Includes bibliographical references and index.
    ISBN 0-471-00839-7 (acid-free)
    1. Regression analysis—Graphic methods—Data processing.
  I. Weisberg, Sanford, 1947-   . II. Title. III. Series.
  QA278.2.C663   1994
  519.5'36'078—dc20   94-19291

Printed in the United States of America

10 9 8 7 6 5 4

To Jami, Jason and Christopher

R. D. C.

To Carol and Stephanie

S. W.

Contents

Preface

This book is about using graphs to understand how a response variable depends on one or more predictors, bringing two new sets of ideas to this old problem. First, we now have the potential to learn much more from graphs because high-quality computers allow us to draw graphs easily, to interact with them, and to use motion to convey information. Second, foundational guidance for graphics is available from an emerging theory that tells us what to look for in graphs, and how to interpret what we see. Combining these two advances gives us a new field of *regression graphics*. We describe in this book how to discover the structure of most regression problems, and illustrate how to translate that structure into useful models when appropriate.

We have written this book to be accessible to students who are currently learning linear regression for the first time. *This is not a research monograph*, although some of the material contained here cannot be found elsewhere. We have taught several times using preliminary versions of this book, integrating the material into a single-quarter course that formerly covered just linear regression for upper division undergraduate and beginning graduate students in a variety of fields. This new methodology complements standard methods and does not replace them.

The *R-code*

Many of the ideas described in this book require an approach to statistical computing that is not currently available in standard statistical software, so we have included software with this book. The program is called the *R-code*, short for regression code. It is an application written in Luke Tierney's *Xlisp-Stat* language. The disks included in the back of this book, for both Macintosh and Windows computers, contain both the *R-code* and

Xlisp-Stat. However, programming in *Xlisp-Stat* or any other language is not required to use the *R-code*.

The *R-code* is a nearly complete system for linear regression analysis, and could be used as the only program in a linear regression course, even without using its graphics capabilities. This book is the manual for the *R-code*, although most of the ideas in the book are independent of the program.

Outline and Summary

Because much of the material in this book is novel, we present here an overview of our general approach and philosophy. This overview is at a higher mathematical level than the book itself.

Statistical graphics can be viewed as consisting of two broad types of tasks: construction and interpretation. Construction refers to everything involved in the production of a graphical display. In this book we rely on histograms, 2D scatterplots, 3D scatterplots and scatterplot matrices, along with various interactive graphical enhancements such as smoothing, linking, brushing, detrending, projecting and orthogonalizing. The first five chapters are devoted to introducing the graphical tools and showing how they relate to regression problems. Most *plot controls*, which are plotting tools that allow the user to interact with graphs, are introduced in these chapters. Detailed explanation of a few plots controls, such as the ability to delete selected points from an analysis interactively, is reserved until later chapters.

In Chapter 6 we begin discussion of interpretation, which refers to the way graphical displays are characterized and how those characterizations might be used to form conclusions about the regression problem itself or the continuation of the analysis. Useful interpretations are possible when the graph is fundamentally connected to the underlying regression problem.

The primary graphical goal of this book is to form displays that can provide visual information on how the distribution $F(y|x)$ of a univariate response y given a $p \times 1$ predictor x changes with the value of x. We emphasize two important characteristics of the conditional distribution, the regression function $E(y|x)$ and the variance function $var(y|x)$, but for this description the more general goal is useful. When there is only one predictor, the conditional distributions $F(y|x)$ can be studied with a 2D scatterplot of the data, with the response on the vertical axis and the predictor on the horizontal axis. The basic ideas for this are introduced in Chapters 2 and 3. Even in this relatively simple setting, unfamiliar though intuitive ideas are required. When $p \geq 2$, we might think in principle of drawing a $(p+1)$-dimensional plot, with the response on the vertical axis, and the predictors on p "horizontal" axes. Of course if $p \geq 3$ this plot is only theoretical in our three-dimensional world, so we need additional structure to guide the construction of practically usable plots.

A fundamental idea we use is *dimension reduction* (see Li, 1991 and Cook, 1994). We want to replace x with the smallest number $d \leq p$ of linear combinations $(h_1^T x, \ldots . h_d^T x)$ of x without loss of information on $F(y|x)$. Let $\eta = (h_1, \ldots, h_d)$. The condition that x can be replaced by $\eta^T x$ without loss of information is equivalent to the restriction that

$$F(y|x) = F(y|\eta^T x) \qquad (0.1)$$

for all values of x in the relevant sample space. If (0.1) holds then it also holds with η replaced by ηA where A is any full-rank $d \times d$ matrix. Thus (0.1) is really a statement about $S_{y|x}(\eta)$, the d-dimensional subspace of $\Re^p$ spanned by the columns of η, rather than about any particular basis η. The subspace $S_{y|x}$ is called the *minimum dimension-reduction subspace* for $F(y|x)$ in Cook (1994). In this book the dimension $d = \dim(S_{y|x})$ of $S_{y|x}$ is called the *structural dimension* of the regression. We generally restrict attention to regression problems with $d \leq 2$.

If $S_{y|x}$ were known and $d \leq 2$, then a plot with y on the vertical axis and $\eta^T x$ on the horizontal axis or axes would contain all sample information on $F(y|x)$. Such plots are called *ideal summary plots*. When $d = 0$, so that $F(y|x) = F(y)$, a histogram of y is an ideal summary plot. If $d = 1$ then $F(y|x) = F(y|h_1^T x)$, and the 2D scatterplot of y versus $h_1^T x$ is an ideal summary plot. A three-dimension ideal summary plot is required if $d = 2$. The estimation of ideal summary plots is a theme of this book.

As an example, the usual homoscedastic linear regression model with independent errors,

$$y|x = \beta_0 + \beta^T x + \varepsilon \qquad (0.2)$$

has structural dimension 1 with $S_{y|x} = S(\beta)$. This model also restricts the regression function to be linear in x and the variance function to be constant. For any nonzero constant c, a plot of y versus $c\beta^T x$ is an ideal summary plot. In Chapter 6 we restrict attention to model (0.2) with $p = 2$ and describe how to use a 3D plot to obtain a visual estimate of an ideal summary plot. We also introduce the notion that the distribution of the *predictors* is central to graphical analyses, a topic that becomes more important as the book progresses. Predictors x are called *linear predictors* if $E(x|B^T x)$ is a linear function of the value of $B^T x$ for all conformable matrices B.

In the first three sections of Chapter 7 we describe how 3D plots can be used to construct visual estimates of $S_{y|x}$ when $p = 2$, but without assuming a particular model. The approach we take stands in contrast to the more usual approach of fitting a target model such as (0.2), and then using the residuals from the fit to criticize the model with the eventual goal of model refinement (see Cook and Weisberg 1982, Sec. 1.2). The discussion of many-predictor problems is started in Section 7.4 by restricting consideration to regressions without a model but with structural dimension 1. There we rely on a key result of Li and Duan (1989) to justify using the ordinary least squares estimate of β from (0.2) as a basis for estimating $S_{y|x}$ when the predictors are linear, even though model (0.2) may clearly be wrong. In Chapter 8 we use sliced inverse regression (Li, 1991) and a related graphical procedure to obtain estimates of $S_{y|x}$ without restriction on the number of predictors or the structural dimension, but with linear predictors required.

In Chapter 9 we discuss the use of coordinate-wise predictor transformations to simplify a regression problem. Let $t(x) = (t_j(x_j))$ denote the $p \times 1$ vector of transformed predictors. Often transformations can be found so that $F(y|x) = F(y|t(x))$ and $\dim(S_{y|t(x)}) < \dim(S_{y|x})$, thereby forming a related regression with reduced structural dimension. The approach of Chapter 9 is restricted to coordinate-wise transformations via generalized additive models and component-plus-residual plots. Transformations t_j are estimated interactively and graphically using plot controls to extract smooths. The discussion ends with one iteration of a basic backfitting procedure, which is often adequate with linear predictors.

Response transformations are useful for inducing a linear regression function in problems with structural dimension 1. A graphical method for selecting linearizing response transformations is described in Chapter 10. Standard Box-Cox methods are discussed as well.

In Chapter 11 we turn to diagnostic residual plots assuming that a target linear model has been developed, perhaps using the graphical methods of the previous chapters. Let e denote population residuals from the target model. The general diagnostic issue is to determine graphically if there is information in the data to contradict the conjecture that $\dim(S_{e|x}) = 0$, which is equivalent to the conjecture that the target model is adequate. We make use of linear predictors to reduce the work involved. The treatment of outliers and influential observations in Chapter 13 is fairly standard, and this chapter is devoted mainly to showing how unusual observations can be addressed in the *R-code*.

An adequate linear model (0.2) is assumed in Chapters 12 and 14. Graphical methods are suggested for visually assessing the contributions of individual predictors (Chapter 12) and for constructing joint 2D and 3D confidence regions for the coefficients (Chapter 14).

Organization and Style

When writing this book, we envisioned the reader sitting at a computer and reworking the examples in the text. Indeed, some of the text can read as if we were next to the reader, suggesting what to do next. To maintain this low-key style, we avoided technical discussions and didn't spend much time on numerical results. References and technical comments are collected in the Complements section at the end of each chapter.

Teacher's Manual

A teacher's manual, giving solutions to the exercises, our course outline, and other material, is available from the publisher, John Wiley & Sons, Inc.

R-code Questions

Questions concerning the *R-code* and bug reports should be sent via e-mail to rcode@stat.umn.edu.

Acknowledgments

Several friends, colleagues and students have been generous with their time and ideas in working though this material. We would like first to thank Bret Musser and Rob Weiss. Both read this work line by line and suggested many improvements. Bret drew the final versions of all the graphs in this book, while Rob's feedback from his class at UCLA has been very helpful. Paul Alper and a group of his colleagues at the University of St. Thomas worked through the book, and provided important insights to improve the presentation. An earlier version of the manuscript was read by Kyle Matschke, Martyn Smith, and Rob McCulloch, all of whom provided very useful comments. We are also grateful to Steve Taff, Bob Semmes, and many students at the University of Minnesota for their comments and reactions to this work. Luke Tierney, the author of *Xlisp-Stat*, helped resolve key computational issues in the *R-code*, which developed over a period of approximately 6 years.

Throughout this work, we have been supported by the National Science Foundation. The finishing touches on this book, the teacher's manual and all supplementary material were supported in part by the National Science Foundation's Division of Undergraduate Education through grant DUE#93-54678.

R. Dennis Cook
Sanford Weisberg

CHAPTER 1

Getting Started

The primary goal in a regression analysis is to understand how a *response variable* depends on one or more *predictors*. With just one predictor, a two-dimensional (2D) graph with the predictor on the horizontal axis and the response on the vertical axis shows the dependence. With many predictors, a fully graphical approach is harder.

The central theme in this book is that graphs can be used to visualize the dependence of a response on predictors, even in high dimensions. For this to be feasible, we need to develop a new vocabulary to describe graphs and new tools to help extract information from them. The paradigm of characterizing the contents of a graph and then using the results to guide subsequent analysis recurs in every chapter of this book.

Many of the graphical tools we use are *kinetic* or *interactive*. Kinetic plots use motion on the computer screen to convey information. A rotating three-dimensional (3D) plot is an example of a kinetic plot because it uses motion to create the illusion of a rotating point cloud in 3D. These plots have the potential to increase our understanding of a regression problem, since we may be able to find dependence that is not visible in a 2D plot. The types of dependence that can be found in 3D plots and methods for using 3D plots are topics in later chapters.

Interactive graphics allow the user to change the appearance of a graph, by changing its shape or size, marking or coloring points, adding or removing points, linking a graph with others, or adding additional visual

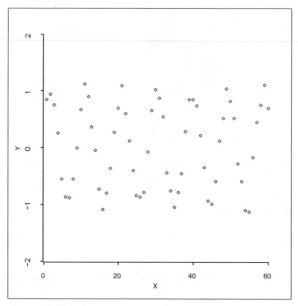

Figure 1.1. What do you see in this scatterplot?

cues. Consider, as an example, the data displayed in Figure 1.1. The best graphics let the user discern patterns in a plot. Are any patterns apparent here? After studying this plot for a while, turn the page and compare it to Figure 1.2. The data are the same in the two plots. The only difference is in the *aspect ratio*, the physical length of the vertical axis divided by that of the horizontal axis. A sine curve with a small amount of error is plainly evident in the second plot. Few people find this pattern in the first plot. While this example is extremely simple, the lesson is very general: to find patterns or dependence in a graph, our eye often needs visual aids like changing the aspect ratio. We will encounter many examples to reinforce this lesson in later chapters.

This book is about using graphs in regression analysis. It includes an associated computer program called the *R-code*. Readers of the book can, and should, reproduce almost all the graphs using the *R-code* with their own computer. Section A.2 in the Appendix gives simple instructions for installing the *R-code*. The *R-code* is written in the *Xlisp-Stat* language, but you don't need to know anything about *Xlisp-Stat* or about computer programming to use the *R-code*.

1.1   DOING THE EXAMPLES

Almost all computing with the *R-code* requires only mouse input and typing items in dialogs. The usual procedure is

1. Load the *R-code* as described in Section A.2. On the Macintosh, a window will be created called *Xlisp-Stat*. With Windows, two new windows are created, one called *Xlisp-Stat* and one called *Listener*. *Listener* is the same as the *Xlisp-Stat* window on the Macintosh. We will call this the *text window*, since all printed input and output appears here.[1] After you get familiar with the program, you may want to move the text window to a corner of your screen and perhaps resize it. The prompt > will appear in the text window when the program is loaded.

2. Load a file containing either a data set or a demonstration program. This requires selecting "Load" from the "File" menu[2] and then choosing the name of the file you want to load.

3. If the file is a demonstration, a menu will be created. Activities described in this book are obtained by selecting items from the menu. If the file is a data file, you will get the regression dialog described in Section 1.3. Most features of the program are described in the main part of this book, while a few items are discussed in the Appendix. Use the index to find the topic you want.

Using the demonstration and data files, you will be able to draw almost every figure in this book, plus a lot more.

1.2   A VERY BRIEF INTRODUCTION TO *Xlisp-Stat*

We begin with a brief introduction to *Xlisp-Stat* by using an example to illustrate some basic numerical and graphical methods. You will be typing commands in this example, even though typing plays a minor role in this book. Since *Xlisp-Stat* is a language, you can use it to write programs and functions; the *R-code* was written in this way. Learning to program is not necessary to use the *R-code*, however.

---

[1]On Unix workstations, the text window is the window you used to start *Xlisp-Stat*.

[2]On Unix workstations there is no file menu, so a file is loaded by typing in a command. For example, to load the file `wool.lsp` in the directory `R-data`, type in the command `(load "R-data/wool")`. The extension `.lsp` need not be specified. There are special functions for loading the data and demonstrations in the `R-data` directory. To load a data file, you can type `(load-rdata "wool")`. To load the demonstration `demo-2d.lsp`, you can type `(load-rdata "demo-2d")`.

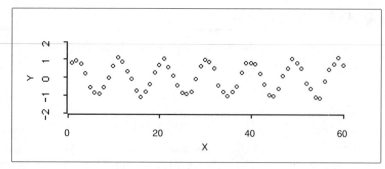

Figure 1.2. Same data as in Figure 1.1, except the aspect ratio is changed from about 1:1 to about 1:5.

The data for our example contain two variables: (1) cigarette consumption per person per year in 1930 and (2) 1950 male lung cancer death rate per million males. The variables are measured for the United States, Canada, and nine countries in Europe. One possible analysis goal for these data is to see if lung cancer death rate is associated with cigarette smoking. The smoking data lead the cancer data by 20 years to allow time for cancer to develop.

1.2.1 Entering Data

Enter the data into your computer as shown below. Duplicate the commands on each line following the prompt >. The number of items you put on one line is arbitrary. Each command you type ends by hitting the Return key. The computer will respond by providing a new prompt when a command is finished.

```
> (def cancer-rate (list 58 90 115 150 165 170 190 245 250
        350 465))
CANCER-RATE
> (def cig-consumption (list 220 250 310 510 380 455 1280 460
        530 1115 1145))
CIG-CONSUMPTION
> (def country (list "Iceland" "Norway" "Sweden" "Canada"
                    "Denmark" "Austria" "USA" "Holland"
                    "Switzerland" "Finland" "Great_Britain"))
COUNTRY
```

You have now created three lists of data with the names *cancer-rate*, *cig-consumption*, and *country*. To understand a little about how this happened, it will probably be helpful to know that *lisp* stands for *list* *processor* and, as the name implies, the language works by processing lists. A list is just a series of items separated by spaces and enclosed in parentheses. The items in a list can themselves be lists, or even more complicated constructions. Generally, the first item of a list is an instruction that tells the program what to do with the rest of the list. For example, the list (+ 1 2) contains three items, +, 1, and 2, and it instructs the program to add 1 and 2.

Let's now return to the first of the three commands you just typed. The main list contains three items, def, cancer-rate, and (list 58 90 115 150 165 170 190 245 250 350 465). The first item, def, is the instruction that tells the program you want to define something. The next item, cancer-rate, is the name to be defined, and the final item is the definition. The definition is itself a list consisting of 12 items: list and the numbers 58, ..., 465. The first item is the instruction for forming a list. In sum, the first command that you just typed instructs the program to define cancer-rate to be a list of 11 numbers. After you pressed the return key, the computer responded with the name of the list and then the next prompt.

The names of lists or other objects in *Xlisp-Stat* can be of any length but must not contain spaces. A few names like pi, t, and nil are reserved by the system. If you try to use one of these, you will get an error message. You will rarely get error messages from the *R-code*, but some possible messages and their likely causes are given in Section A.8.

Remember these points when typing information into *Xlisp-Stat*:

- The parentheses must match. In both the Macintosh and Windows versions, *Xlisp-Stat* flashes matching parentheses to make this easier.
- Quotation marks must match. If they do not match, the program will get very confused, and it will never return you to the > prompt or give you any printed output.
- You can have any number of items on a line.
- A command can actually take up several lines on the computer screen.
- Each command ends with a Return after the final right parenthesis.
- Items in a list are separated by white space, consisting of either blanks or tabs.
- If the computer does not respond to your typing, you need to escape from the current command to start over. On the Macintosh, you can select the item "Top Level" from the "Command" menu or press the command (cloverleaf) and period keys at the same time. On other systems, press Control-C.
- If you have an error, you can use standard cut-and-paste methods to edit the command for reentry. The way this works is most easily learned by experimenting.

1.2.2   Working with Lists

You can do many things with lists. The simplest is just to display them. To
see the list cig-consumption, type

```
> cig-consumption
(220 250 310 510 380 455 1280 460 530 1115 1145)
```

Don't forget the return after cig-consumption. To get the number of
items in a list, use the length function:

```
> (length cig-consumption)
11
```

If you do not get 11, then you have incorrectly entered the data: try to find
your error, and then retype or use the cut-and-paste features of the program
to correct the data before continuing.

To get the mean and standard deviation of *cig-consumption*, you can use
the mean and standard-deviation functions:

```
> (def xbar (mean cig-consumption))
XBAR
> xbar
605
> (def sdx (standard-deviation cig-consumption))
SDX
> sdx
384.37
```

In each of these statements, a value is computed and stored as a constant
so it can be used in a later calculation. If you do not plan to use the value
later, you need not store it. For example, to get the natural logarithm of
cig-consumption, type

```
> (log cig-consumption)
(5.39363 5.52146 5.73657 6.23441 5.94017 6.1203 7.15462
6.13123 6.27288 7.01661 7.04316)
```

while the expression

```
> (exp (log cig-consumption))
(220 250 310 510 380 455 1280 460 530 1115 1145)
```

exponentiates the natural log of `cig-consumption`, which simply returns the original values. The function `log` can take two arguments, a value and a base. For example, (`log 4 10`) will return the logarithm of 4 to the base 10, while (`log cig-consumption 2`) will return a list of the logarithms to the base 2 of the elements of `cig-consumption`. If the second argument is missing, then natural logarithms are used. The *R-code* uses natural logarithms.

Xlisp-Stat can be used for simple calculations. For example,

```
>  (+ 2 4 5 6)
17
>  (- 17 2 3)
12
>   (* 3 xbar)
1815
> (/ cig-consumption 12)
(18.3333 20.8333 25.8333 42.5 31.6667 37.9167 106.667 38.3333
44.1667 92.9167 95.4167)
```

These statements illustrate four basic operations that can be applied to lists of numbers. The general form of the operation may seem a bit unnatural: after the opening parenthesis comes the instruction and then the numbers. For addition and multiplication, this does the obvious thing of adding all the numbers and multiplying all the numbers, respectively. For subtraction and division, it is not necessarily obvious what happens with more than two arguments. As illustrated above, (`- 17 2 3`) subtracts 2 from 17 and then subtracts 3 from the result. Similarly, (`/ 8 2 4`) divides 8 by 2 and then divides the result by 4, giving an answer of 1.

The four basic arithmetic operations can be applied to pairs of lists having the same length:

```
> (* cancer-rate cig-consumption)
(12760 22500 35650 76500 62700 77350 243200 112700 132500
390250 532425)
```

The operation is applied to the lists elementwise, resulting in a new list having the same length as the original two lists.

The exponentiation function is illustrated by

```
> (^ 3 4)
81
```

which gives the value of $3^4 = 81$.

1.2.3   Calculating the Slope and Intercept

You can also do more complicated computations. Let x_i refer to the ith value of the predictor *cig-consumption* and let y_i refer to the ith value of the response *cancer-rate*. To get $SXY = \sum(x_i - \bar{x})(y_i - \bar{y})$ and $SXX = \sum(x_i - \bar{x})^2$, you can type:

```
> (def ybar (mean cancer-rate))
YBAR
> (def sxy (sum (* (- cig-consumption xbar)
                   (- cancer-rate ybar))))
SXY
> sxy
338495
> (def sxx (sum (^ (- cig-consumption xbar) 2)))
SXX
> sxx
1.4774e+06
```

The hard part in this calculation is getting the parentheses to match and remembering that functions like + or * or `log` go at the start of an expression. *Xlisp-Stat* allows mixing lists and numbers in one expression. In the above example, `(- cig-consumption xbar)` will subtract `xbar`, a number, from each element of the list `cig-consumption`.

To complete the calculations of the slope and intercept obtained by ordinary least squares regression of *cancer-rate* on *cig-consumption*, use the usual formulae:

```
> (def b1  (/ sxy sxx))
B1
> (def b0 (- ybar (* b1 xbar)))
B0
> (list b0 b1)
(65.7489 0.229115)
```

The estimated intercept is thus about 65.75 and the estimated slope is about 0.23. We will use these values a bit later to add the fitted line to a scatterplot of *cancer-rate* versus *cig-consumption*.

1.2.4   Drawing a Histogram

We now turn to some simple graphics. The *R-code* is designed to make graphics easy, but for now we will use the built-in functions in *Xlisp-Stat*.

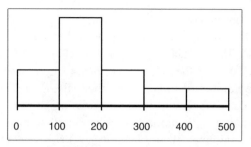

Figure 1.3. Histogram, with five bins.

To get a histogram of the values of *cancer-rate*, type the command

```
> (histogram cancer-rate)
```

The histogram, as shown in Figure 1.3, appears in a separate window on the screen called a *graphics window*. The number of graphics windows opened at one time is limited only by computer memory. The graphics window can be moved, resized, or dismissed using standard mouse movements:

- To move a window, hold down a mouse button in the top margin of the plot, called the *title bar*, and drag the plot around your screen. When you let go of the mouse button, the window stays in its new location.

- To resize a window, hold down the mouse button in the lower right corner of the window. As the mouse is moved, the lower right corner of the plot is moved as well.

- To dismiss a window on the Macintosh, push the mouse button in the small square in the upper left corner of the window. With Windows, push the mouse button in the upper left corner of the window and then select "Close" from the pop-up menu.[3]

In *Xlisp-Stat*, all graphics windows have a *menu*. Make sure that your histogram is the *front window* by clicking the mouse button anywhere in this window or, with Windows, selecting the title of the window from the "Windows" menu. You will now have an additional menu called "Histogram" in the menu bar. If you have many windows on the screen, only the menu for the front window is displayed.[4]

---

[3]On Unix systems, push the button marked "Close" on the plot.
[4]On Unix systems, the menu is attached to a button called "Menu" at the upper right of the graph.

From the histogram you have just drawn, the data are seen to lie between about 0 and 500. A single mode is apparent between 100 and 200, with more countries with rates above the mode than below the mode, so the data appear to be skewed to the right. The visual impression of a histogram depends on the number of bins. Using the "Histogram" menu, select the item "Change Bins." You will then get a dialog. In the text area in the middle of this dialog is the number 5, indicating that the plot currently has 5 bins. Type the number 10 in place of 5.[5] Press the "OK" button, and the number of bins changes. With 10 bins, the pattern that was initially clear is now more difficult to see. Of course this small data set has only 11 observations, so 10 bins is certainly too many. The *R-code* has an easier way of changing the number of bins, to be illustrated later in this chapter.

1.2.5   Drawing a Scatterplot

Without dismissing the histogram, create a scatterplot as follows: Click the mouse in the text window to make it the front window. A scatterplot with *cig-consumption* on the horizontal axis and *cancer-rate* on the vertical axis is created by typing the command

```
> (def p (plot-points cig-consumption cancer-rate))
```

This will draw a graph and name it p. You can modify the plot by sending *messages* to p. After clicking the mouse again in the text window to make it the front window, type

```
> (send p :add-labels "cig-consumption" "cancer-rate")
```

This sends the graph p the :add-labels message, telling the plot to use the arguments "cig-consumption" and "cancer-rate" as the labels. If p were a histogram, the :add-labels message would use only one argument.

As a second example of sending messages, recall that the list country contains the names of the countries. Type

```
> (send p :point-label (iseq 11) country)
```

The message :point-label is used to associate labels with points in the plot. It has two arguments, a list of case numbers, and a list of labels. The function (iseq 11) returns the sequence of integers (0 1 2 3

---

[5]On Unix workstations, the cursor may need to be placed in a dialog before you can type in it.

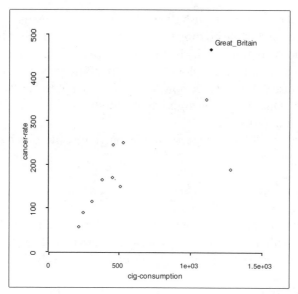

Figure 1.4. Scatterplot of 1930 cigarette consumption against 1950 lung cancer death rate for 11 countries.

4 5 6 7 8 9 10), which are the case numbers. This list starts at
0, since *Xlisp-Stat* always numbers elements in a list starting with 0. The
message associates the 11 country names with the 11 points in the plot.
The plot on the computer screen is not immediately changed after sending
the :point-label message. To see the consequences of sending this
message, first make sure your scatterplot is the front window. The name
of the menu for a scatterplot is "Plot." Select the "Show Labels" item from
this menu. Selected points and their labels will be printed on the graph.
A point can be selected by clicking the mouse button on it. More points
can be selected by holding down the shift key and clicking on the points.
Several points can be selected at the same time by enclosing them within
the selection rectangle that appears while holding down the mouse button
and dragging.

   Figure 1.4 shows that cancer rate generally increases with cigarette consumption. The variability of cancer rate seems to increase as well. What
are the three countries corresponding to the three largest cigarette con-
sumptions? One of the three points, the one with the highest cancer rate,
is highlighted in Figure 1.4. What is the country that seems least well fit
by an increasing relationship between consumption and cancer rate?

   From the scatterplot's menu, select the item "Link View." Move the
scatterplot so both it and the histogram of *cig-consumption* can be seen at the
same time. Click on the histogram, and from its menu select "Link View"

as well. In the histogram, push down the mouse button slightly above the rightmost bar, and while holding the button drag down to select the bar. Let go of the mouse button. In the histogram, your selection is highlighted by turning black. In the scatterplot, the corresponding points are highlighted as well because these plots are now *linked*. Linking plots can be a powerful tool in understanding data by highlighting points corresponding to the same cases in several plots.

You can draw the fitted regression line on your scatterplot by typing the message

```
> (send p :abline b0 b1)
```

where b0 b1 are the names for the estimates that you computed in Section 1.2.3. This method can be used to add any line to a plot. The two arguments to :abline specify the intercept and slope, respectively, of the line to be drawn.

To get a plot of {fitted values, residuals}, where the axes are listed in the order {horizontal axis, vertical axis}, compute the fitted values and the residuals, and then draw the plot:

```
> (def fit-values (+ b0 (* b1 cig-consumption)))
FIT-VALUES
> (def residuals (- cancer-rate fit-values))
RESIDUALS
> (def p1 (plot-points fit-values residuals))
P1
```

The :add-labels message can be used to add axis labels to the plot.

1.2.6   Saving and Printing Text

Xlisp-Stat does not have a built-in method for printing. Output from the program must be saved to a file and ultimately printed by some other program, such as a word processor or art program. Two techniques are available for saving printed output. The simplest is to cut and paste: select the text you want to save by dragging the mouse across it with the button down; copy it to the clipboard; switch to another application, such as a word processor, and paste the output into the application. In a word processor, you should use a fixed-width font like Courier or Monaco, or else columns won't line up properly.

The second method saves *all* the results in the text window to a file. Saving is started by selecting the item "Dribble" from the "Data" menu on the Macintosh or the "File" menu under Windows.[6] Choose a file name using a standard dialog. All text will be put in this file until you select "Dribble" a second time. The resulting plain text file can then be read by any word processor or editor. You cannot toggle the "Dribble" file on and off. If you select "Dribble" a third time with the same file name, the file will be overwritten without warning.

1.2.7   Saving and Printing a Graph

On the Macintosh and Windows versions, a graph is saved by transferring it to a document for another application. Make sure the graph you want is the front window. Then, select "Copy" from the "Edit" menu. Switch to the other application, and paste the graph into a document by selecting "Paste" from that program's "Edit" menu.[7]

1.2.8   Quitting *Xlisp-Stat*

To quit from *Xlisp-Stat* on the Macintosh, select "Quit" from the "File" menu. Using Windows, select "Exit" from the "File" menu. When using any system, you can also type the command `(exit)`, followed by a return, in the text window. On Unix systems, Control-D also quits the program.

1.3   AN INTRODUCTION TO THE *R-code*

So far we have used standard *Xlisp-Stat* commands to get results. For most of the book, we will use the *R-code*.

The file `cancer.lsp` gives the data used in the cancer rate example. This file includes the data and some documentation and code that will start the *R-code*. The file is located in the directory or folder called `R-data`. The file can be loaded by selecting "Load" from the "File" menu, selecting the directory `R-data`, scrolling until you see the file name `cancer.lsp`, and then double clicking on the name.[8] This will print the documentation, read in the data, and assign names to the data. You will then get a dialog similar to Figure 1.5, which is the *standard regression dialog*.

---

[6]On Unix workstations, use the command `(dribble "filename")`, where `filename` is any file name you like.

[7]On Unix workstations, plots can be saved as a PostScript bitmap by selecting the "Save to File" item from the plot's menu.

[8]With Windows, if the list of available files in the file selection dialog does not include the file you want, type `*.lsp` in the text area at the top of the dialog, and then press Open. This will return you to the file selection dialog, but all files on your system that have the `.lsp` extension will be accessible.

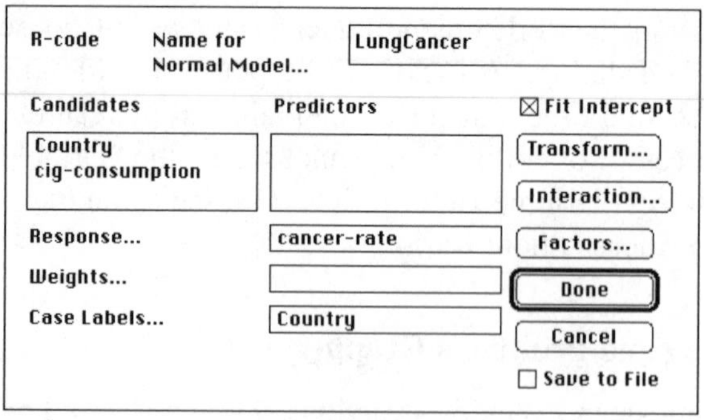

```
┌────────────────────────────────────────────────────────┐
│  R-code      Name for        ┌──────────────────────┐   │
│              Normal Model... │ LungCancer           │   │
│                              └──────────────────────┘   │
│  Candidates        Predictors          ⊠ Fit Intercept  │
│  ┌──────────────┐ ┌──────────────┐    ┌─────────────┐   │
│  │ Country      │ │              │    │ Transform...│   │
│  │ cig-consumption│ │              │    └─────────────┘   │
│  │              │ │              │    ┌─────────────┐   │
│  └──────────────┘ │              │    │ Interaction.│   │
│                   └──────────────┘    └─────────────┘   │
│  Response...      ┌──────────────┐    ┌─────────────┐   │
│                   │ cancer-rate  │    │ Factors...  │   │
│  Weights...       └──────────────┘    └─────────────┘   │
│                   ┌──────────────┐    ┌─────────────┐   │
│  Case Labels...   │ Country      │    │   Done      │   │
│                   └──────────────┘    └─────────────┘   │
│                                       ┌─────────────┐   │
│                                       │   Cancel    │   │
│                                       └─────────────┘   │
│                                       □ Save to File    │
└────────────────────────────────────────────────────────┘
```

Figure 1.5. The standard regression dialog.

Initially, the "Candidates" at the left of the regression dialog include all the variables in the data set. As shown in the figure, *cancer-rate* has been specified as the response. This was done by clicking the mouse once on the name *cancer-rate*, then moving the mouse to the empty box for "Response," and clicking once again. Double click on *cig-consumption* and it moves to the "Predictors" window. The variable *Country* appears in the "Case Labels. . . " box; the *R-code* will use a list of text for labels if one is provided in the data. To move a variable back to the "Candidates" box, double click on its name. At the top of the dialog you can specify a name for the model; the default name is "LungCancer," which can be changed by typing a different name. When finished, push the "Done" button.

Starting from Figure 1.5, double click on *cig-consumption* and then "Done" to finish the dialog. The following will then happen:

1. The linear regression of *cancer-rate* on *cig-consumption* will be calculated and summary statistics will be printed in the text window.
2. All the columns of data in the file will become lists of numbers that you can use in arithmetic computations. For example, typing `(/ cancer-rate cig-consumption)` will print the ratio of these two quantities.
3. A menu will be created with the name "LungCancer" or whatever you typed in for the name of the regression.
4. Finally, a *regression object* will be created with the same name as the name on the menu. Messages to regression objects can be sent just as we did to plots.

You can now study the regression further by using the items in the *regression menu*, which is called "LungCancer" for this example. The items in this menu are explained throughout the book.

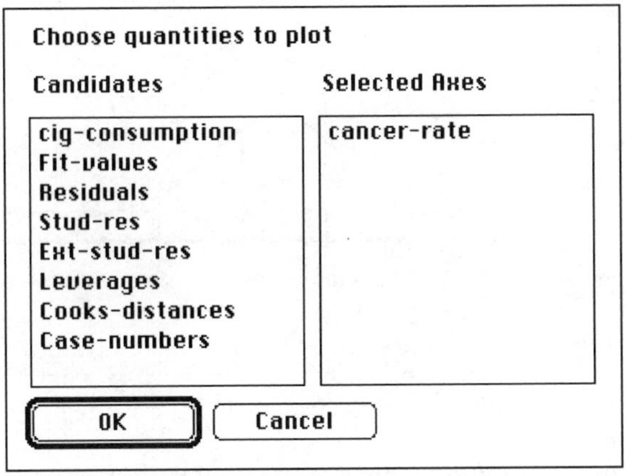

```
┌──────────────────────────────────────────────────────┐
│  Choose quantities to plot                             │
│                                                        │
│  Candidates                    Selected Axes           │
│  ┌────────────────────────┐    ┌──────────────────┐    │
│  │ cig-consumption        │    │ cancer-rate      │    │
│  │ Fit-values             │    │                  │    │
│  │ Residuals              │    │                  │    │
│  │ Stud-res               │    │                  │    │
│  │ Ext-stud-res           │    │                  │    │
│  │ Leverages              │    │                  │    │
│  │ Cooks-distances        │    │                  │    │
│  │ Case-numbers           │    │                  │    │
│  └────────────────────────┘    └──────────────────┘    │
│                                                        │
│  ┌──────────────┐   ┌──────────────┐                   │
│  │     OK       │   │   Cancel     │                   │
│  └──────────────┘   └──────────────┘                   │
└──────────────────────────────────────────────────────┘
```

Figure 1.6. The "Plot of. . ." dialog.

Let's reproduce the histogram constructed earlier. The item "Plot of. . ." in the regression menu is used to draw plots. When you select this item, you will get a dialog like the one shown in Figure 1.6. The left list gives candidates for plotting. The right list gives the selected quantities to be plotted. With many variables, you may need to use the *scroll bar* at the right of the candidates list to see all the entries. Double clicking on a variable name moves it between the lists. As shown in the figure, *cancer-rate* will be plotted. If you push the "OK" button, you will get a histogram of *cancer-rate*. Placing two items in the right list gives a scatterplot. The first variable in the list is assigned to the horizontal axis, and the second item is assigned to the vertical axis. Selecting more than two items gives higher dimensional plots discussed in later chapters.

Use the "Plot of. . ." item to get a histogram of *cancer-rate*, as shown in Figure 1.7. This plot differs from the one you drew earlier in several respects. The most obvious difference is size: the histogram drawn by the *R-code* is larger. Also, the histogram includes several *plot controls*, which are buttons and slide bars that appear to the left, and sometimes below, a figure. Plot controls play a central role in the graphical methodology developed in this book. A histogram has four plot controls. The first control is a slide bar marked "NumBins." Push the mouse button down in the bar. As you hold down the mouse button, the number of bins changes, as given by the number above the bar. In the figure shown, the number of bins is five. The second slide bar is called "GaussKerDens." This control fits a smooth curve to the histogram. The smoothness of the curve fit depends on a *window width* parameter, and this is chosen in the slide bar. The larger the value of the window width, the smoother the curve. The curve shown

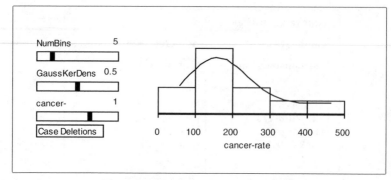

Figure 1.7. *R-code* histogram of *cancer-rate*.

is very smooth, representing a distribution that is skewed to the right with a few large values.

The next slide bar is used to replace the data in the plot by a power transformation of the data, with the power selected in the slide bar; a full discussion of this slide bar is deferred to Section 3.2. The "Case Deletions" item creates a *pop-up menu* to delete cases from a data set. The use of this plot control is illustrated throughout the book.

To redraw the residual plot mentioned at the end of Section 1.2.5, select "Plot of. . ." from the regression menu, and then double click first on *Fit-values* and then on *Residuals* and then push the "OK" button. In the plot's menu, select the "Show Labels" item, and then select the point in the lower right of the graph. The United States had very high cigarette consumption but low cancer rate. You might try deleting USA from the fit by pushing and holding the box marked "Case Deletions," still with the USA point selected. This will give a pop-up menu; in this menu, select the item "Delete Selection from Regression." This will delete the point and cause the regression to be recomputed and the plot to be updated with new residuals and fitted values. Select "Display Fit" from the "LungCancer" menu to get the new fitted equation. Another item in the "Case Deletions" pop-up menu can be used to return USA to the model. You may need to use the "Rescale Plot" item from the plot's menu to see all the points on the plot at one time.

1.4   USING YOUR OWN DATA

The *R-code* can read data from a plain text file containing only data or from a special file that starts the *R-code* automatically.

Let's begin with reading a file containing only data. One file of this type is listed in Table 1.1. This file has 13 rows and 5 columns. Each row

Table 1.1. A Sample Data File

7	26	6	60	78.5
1	29	15	52	74.3
11	56	8	20	104.3
11	31	8	47	87.6
7	52	6	33	95.9
11	55	9	22	109.2
3	71	17	6	102.7
1	31	22	44	72.5
2	54	18	22	93.1
21	47	4	26	115.9
1	40	23	34	83.8
11	66	9	12	113.3
10	68	8	12	109.4

in the file gives all the data for one particular case. Each column gives
the values for one variable. As shown in the table, the data line up one
beneath the next, but this is not necessary. All that is required is that (1) the
same number of values appear on each row of the file and (2) the values
are separated by white space, either blanks or tabs. Columns of text are
permitted, where text is any collection of nonblank characters that is not
interpretable as a number. Missing values are not permitted.

A duplicate of the table is given in the file `halddata.lsp` in the
`R-data` folder. On the Macintosh, you can view or edit the file using the
built-in editor in *Xlisp-Stat*. Select the item "Open Edit" from the "File"
menu, and then from the resulting dialog open the `R-data` folder and
select the file `halddata.lsp`. This will open an editable window with
the data file in it; you will see it is identical to Table 1.1. In the Windows
version, there is no built-in editor, but there is a separate application called
LSPEDIT that is similar to the Macintosh editor. You can of course use
any editor or word processor, but all data files must be saved as plain text
files.

A plain data file can be introduced into the *R-code* by typing

```
> (r-code)
```

You will then get a dialog to select a file. Choose the file `halddata.lsp`
in the `R-data` folder. The program will read the data and determine that
the file has 5 columns and 13 rows. Further dialogs will be presented so
you can name the 5 columns. The first dialog is shown in Figure 1.8. The
first column of the file, column zero in *Xlisp-Stat* numbering, starts with
the values 7, 1, 11, 11, 7. You can verify that these are the first five values
in the first column of the file. Type the name X1 for this variable; do not

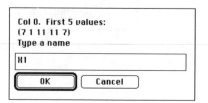

Figure 1.8. Choosing names.

enclose the name in quotation marks. After pushing "OK," you will be prompted for names for the other columns; use X2, X3, X4, and Y. You will then get a standard regression dialog like Figure 1.9. The default name of the regression is *reg*. You might want to change this to something more descriptive.

At this point you can either start regression analysis, save the data and labels to a file, or both. Starting from Figure 1.9, just double click on "X4" and push "Done" to start analysis. In the illustration, the default name for the regression has been changed to the more descriptive "Hald-example." To save the data and labels, push the button "Save to File" in the regression dialog. When you push the "Done" button, you will get another dialog to choose a name for the data file; the program will automatically add the suffix .lsp to the name you choose. A file created this way can be read by the *R-code* using the load command, but it is not in a form that is particularly easy for a human to read.

All of the data files used with this book include both data and a few *R-code* and *Xlisp-Stat* commands. The general format we use for these

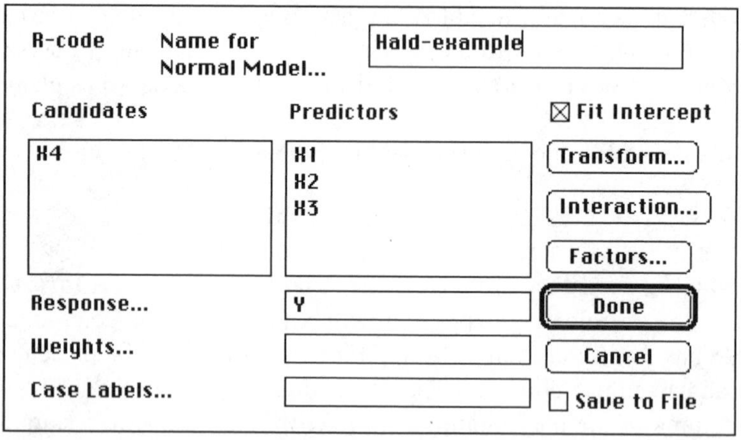

```
R-code      Name for        Hald-example
            Normal Model...

Candidates          Predictors          ☒ Fit Intercept

 X4                  X1                 [ Transform... ]
                     X2
                     X3                 [ Interaction... ]

                                        [  Factors...  ]

Response...          Y                  [   Done   ]

Weights...                              [  Cancel  ]

Case Labels...                          ☐ Save to File
```

Figure 1.9. The standard regression dialog.

Table 1.2. A Data File with Extra Information to Start the *R-code*

```
(defun source () (format t
"Source:  Hald, A. (1960), Statistical Theory with Engineering
Applications, New York:  Wiley, p. 564.
X1, X2, X3 and X4 are percentages of 4
chemicals in the composition of samples of Portland cement.
Y is the heat evolved in calories per gram of cement. ~%"))

(def data (transpose (split-list '(
   7 26   6 60    78.5
   1 29  15 52    74.3
  11 56   8 20  104.3
  11 31   8 47    87.6
   7 52   6 33    95.9
  11 55   9 22  109.2
   3 71  17   6  102.7
   1 31  22 44    72.5
   2 54  18 22    93.1
  21 47   4 26  115.9
   1 40  23 34    83.8
  11 66   9 12  113.3
  10 68   8 12  109.4
  ) 5)))

(def data-names (list "X1" "X2" "X3" "X4" "Y"))
(source)
(r-code :data data :data-names data-names :name "Hald")
```

files is illustrated in Table 1.2, which reproduces the contents of the file hald.lsp in the R-data folder. This standard format for a data file includes (1) a function called source that is used to provide documentation for the data; (2) the data itself, or *Xlisp-Stat* commands that will create the data; (3) labels for all the variables; and (4) a command to execute the function r-code to start regression analysis. The call to r-code has three *keywords*, which are used to pass information to the program. The keyword :data is used to specify the data; :data-names specifies the names of the variables, and :name specifies the initial name for the regression. The argument passed to the :data keyword is also called data, since this is the name used in the file. Unlike many computer languages, *Xlisp-Stat* allows using the same name for several purposes.

If you want to make your own data file of this format, you can modify the file in Table 1.2 using an editor. For the function source, replace everything between the first " and the final ~%")) by text relevant for your data. Then, replace the data shown in the table by your own data.

If your data has 10 columns, replace the `) 5)))` by `) 10)))`. Finally, replace the names for the Hald data by your own names, each name being surrounded by quotation marks. In the call to `r-code`, change `"Hald"` to a name relevant for your data. Save the file as a plain text file with any legal name ending in `.lsp`. The resulting file can be loaded into the *R-code*.

1.5   GETTING HELP

Help is available for many of the functions in *Xlisp-Stat* and for most of the methods in the *R-code* using the functions `help` and `apropos`. These are described in Section A.7.1.

1.6   COMPLEMENTS

The *Xlisp-Stat* language was written by Luke Tierney and is documented in Tierney (1990). Chapter 2 of Tierney's book gives a much more extensive tutorial for *Xlisp-Stat*, which is a particular implementation of the general *Lisp-Stat* language. *Xlisp-Stat* uses the *Xlisp* language written by David Betz. The smoking and lung cancer data are originally from Doll (1955), and are given by Tufte (1974, p. 82). They are also given in the file `cancerdt.lsp` in the `R-data` folder. The data for Exercise 1.2 were furnished by Mike Camden of Wellington Polytechnic.

Several good books on linear regression models have been published in the last decade, and virtually any of these will provide the necessary prerequisite for this book. For fear of leaving one out, we will leave them all out, and leave the choice of book up to the reader.

The slide bar "GaussKerDens" is actually fitting a *kernel density estimate* using a Gaussian kernel. The window width chosen in the slide bar is a fraction of the range in the data. Härdle (1990) and Scott (1992) provide good references for density estimation.

EXERCISES

1.1. Work the examples in Sections 1.2 and 1.3. Turn in the following items: (1) the histogram of *cig-consumption*, first with eight bins and then with four bins; (2) a scatterplot of {*cig-consumption, residuals*}, with the case name of the point corresponding to the largest residual indicated on the plot; (3) the *R-code* output from the "Display fit" item in the regression menu for the regression of *cancer-rate* on *cig-consumption*.

1.2. Data on 56 normal births at a Wellington, New Zealand, hospital are given in file `birthwt.lsp` in the `R-data` folder. The four variables are *BirthWt*, birth weight in grams; *Age*, the mother's age; *Term*, the term of the pregnancy in weeks; and *Sex*, the baby's sex, 0 for girls and 1 for boys.

Use the data file to examine the relationship between the predictor *Age* and the response *BirthWt*. Who tends to have larger babies, older mothers or younger mothers? Can you think of an easy way to identify the points in the plot for girl babies?

CHAPTER 2

Simple Regression Plots

In this chapter we study simple regression problems with *response* variable y and a single univariate *predictor* variable x. The response variable is sometimes called the *dependent* variable, and the predictor variable may be called the *independent* variable, *explanatory* variable, *carrier*, or *covariate*. These names may imply slightly different roles for x, but any distinctions between them are not important here.

A common theme in regression analysis is the characterization of the dependence of y on x as far as possible with the available data. Equivalently, this goal is to characterize how the distribution of y with x held fixed at a particular value, say $\tilde{x}$, changes with $\tilde{x}$. Since this is a fairly important idea in regression graphics, it is helpful to have a convenient way to refer to the response variable y when the predictor x is held fixed. We will use the notation $y|(x = \tilde{x})$ to denote the response variable when the predictor is held fixed at the value $\tilde{x}$. The vertical bar in this notation stands for the word "given." If the particular value $\tilde{x}$ is unimportant for the discussion at hand, we will write more simply $y|x$, understanding that the predictor is held fixed at some value.

Studying all of the ways that the distribution of $y|x$ could change with x can be too unstructured a problem to make much progress. Regression analyses usually focus on characterizing how the true mean of $y|x$, denoted $E(y|x)$, and the true variance of $y|x$, denoted $var(y|x)$, depend on x. The mean $E(y|x)$ is often of primary interest and is called the *regression function*. Similarly, var($y|x$) is called the *variance function*. The regression function is the true mean of y when x is held fixed at a selected value, so it is a function of the value of x. The dependence of E($y|x$) on x and to a lesser extent the dependence of var($y|x$) on x can often be seen in scatterplots. This is the topic of this chapter.

2.1   THINKING ABOUT SCATTERPLOTS

A simple scatterplot with the predictor on the horizontal axis and the response on the vertical axis often provides a very good way to understand how the distribution of $y|x$ changes with the value of x. Our notation for this plot is $\{x, y\}$, where the quantity on the horizontal axis is given first, followed by the quantity on the vertical axis. While this notation is at odds with the common expression of "plotting y versus x," it corresponds to the way graphs are constructed in the *R-code*, by selecting first the quantity to plot on the horizontal axis and then the quantity for the vertical axis.

We begin our discussion of scatterplots with an example. Figure 2.1 gives a scatterplot $\{x, y\}$ of $x = temperature$ in degrees Fahrenheit and of $y = $ daily *ozone* concentration in parts per million, both measured near Upland, in Southern California, for 330 days in 1976. The source of the data does not report if the temperature is the maximum daily temperature or the average temperature. This figure and the others in this chapter can be reproduced by selecting "Load" from the file menu, changing to the folder R-data, and then selecting the file demo-2d.lsp. Figure 2.1 is reproduced by selecting the "Ozone data" item from the resulting "Demo:2D" menu.

Suppose that we use Figure 2.1 to compare the distribution of $y|(x = 40)$ to the distribution of $y|(x = 80)$. There will probably be very few values in the data at which *temperature* is exactly 40 or 80°F. Fortunately, having exact values of $\tilde{x}$ in the data is not necessary to study the distribution of $y|(x = \tilde{x})$. We can use instead all of the values of y that correspond to values of x close to $\tilde{x}$, rather than the values of y corresponding to just those observations where x equals $\tilde{x}$. In graphics terminology this is called *slicing*.

To compare the distributions of $y|(x = 40)$ and $y|(x = 80)$, we slice the plot in Figure 2.1 about the values 40 and 80 on the horizontal axis, and for visual clarity we delete all other points. The resulting plot is shown in Figure 2.2. The portion of the horizontal axis covered by a slice is called the *slice window*, and the width of a slice window is called the *window width*. The two slice windows in Figure 2.2 are centered at 40 and 80, and

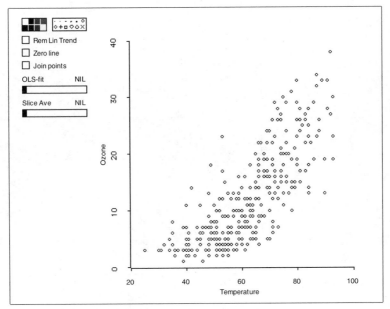

Figure 2.1. Los Angeles ozone data.

the common window width is about 6. The slices in Figure 2.2 are visual aids that enable us to focus on the data in question. Later we will make quantitative use of the slices, but for now it seems clear that the mean of the response for the slice at $x = 40$ is less than the mean for the slice at $x = 80$. The same may be true for the variances as well.

A plot similar to Figure 2.2 can be constructed by first selecting the points that roughly correspond to the slice at $x = 40$ and then, with the Shift key depressed, selecting points that correspond to the slice at $x = 80$. Depressing the Shift key will cause the points selected at $x = 40$ to remain selected while the other points are selected. From the plot's menu choose the item "Focus on Selection." Your plot should now look like Figure 2.2 except perhaps for minor differences in the slices. To return to the plot of Figure 2.1, select the "Show All" item from the plot's menu.

The slices in Figure 2.2 serve as graphical enhancements to aid in the discussion of the statistical interpretation of a scatterplot. They are usually not needed in practice since our eyes provide a smooth transition between the distributions of $y|x$ for adjacent values of x.

2.2   SIMPLE LINEAR REGRESSION

The simple linear regression model is based on two key assumptions about the distribution of $y|x$. To review these assumptions, we begin by writing

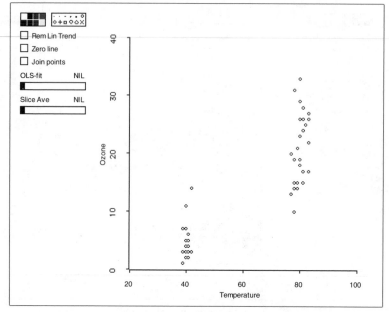

Figure 2.2. Los Angeles ozone data with slices at $x = 40$ and $x = 80$.

$y|x$ as the sum of the regression function and error term,

$$\begin{aligned} y|x &= \mathrm{E}(y|x) + [y|x - \mathrm{E}(y|x)] \\ &= \mathrm{E}(y|x) + \varepsilon|x \end{aligned}$$

where $\varepsilon|x = [y|x - \mathrm{E}(y|x)]$ is the error at x.

The first key assumption in simple linear regression is that the regression function is linear in x,

$$\mathrm{E}(y|x) = \beta_0 + \beta_1 x \tag{2.1}$$

where β_0 and β_1 are the unknown intercept and slope that must be estimated from the data. The second assumption concerns the error term, $\varepsilon|x$. Since $\mathrm{E}(\varepsilon|x) = 0$, this term can depend on x only through second or higher moments. The usual assumption of simple linear regression is that $\varepsilon|x$ does not depend on x at all. Under this second assumption we can write $\varepsilon = \varepsilon|x$ without confusion since the errors do not depend on x. Under this condition the variance function $\mathrm{var}(y|x)$ is just a nonnegative constant, denoted by the symbol σ^2. Later in this book, we will allow the variance function to depend on x.

The simple linear regression model for a particular data set can now be summarized as

$$y|x_i = \beta_0 + \beta_1 x_i + \varepsilon_i \tag{2.2}$$

for $i = 1, 2, \ldots, n$, where n is the total number of observations. The notation $y|x_i$ is shorthand for the response variable when $x = x_i$. We will sometimes use y_i to denote the response variable when $x = x_i$, but we will rely on the fuller notation $y|x_i$ when it seems important for emphasis or clarity of exposition. The errors ε are also assumed to be independent from observation to observation.

Fitting the simple linear regression model (2.2) means obtaining estimates of the parameters. The *R-code* generally does this by using ordinary least squares, which we abbreviate as ols. This leads to a number of quantities that can be useful in graphical investigations. The ols estimates of the regression coefficients β_0 and β_1 are denoted by $\hat{\beta}_0$ and $\hat{\beta}_1$, respectively. The fitted values are denoted by

$$\hat{y}_i = \hat{\beta}_0 + \hat{\beta}_1 x_i \tag{2.3}$$

and the residuals by

$$e_i = y_i - \hat{y}_i \tag{2.4}$$

One of the fundamental advantages of graphical methods is that they allow for an assessment of the assumptions underlying standard regression models. We should always ask if there is information in the data to contradict the linearity assumption or the assumption that errors are independent of the predictor.

2.3 ASSESSING LINEARITY

2.3.1 Superimposing the Fitted Line

One simple method for assessing the linearity assumption (2.1) is to superimpose the ols fitted line onto the scatterplot of the data. Clicking anywhere on the slide bar titled "OLS-fit" in Figure 2.1 will change the word "NIL" above the slide bar to the number 1 and will cause the ols fitted line to appear on the plot of {*temperature, ozone*}. The result is shown in Figure 2.3. Comparing the points in the plot to the ols line, some curvature may be apparent, although this conclusion is far from certain. Clicking anywhere to the left of the *slider*, the filled-in area within the slide bar, will remove the ols line from the plot and return the slide bar to its original state. The same effect can be achieved by dragging the slider to the left.

Clicking more than once on the "OLS-fit" slide bar to the right of the slider or dragging the slider to the right will fit higher order polynomials to the scatterplot. For example, dragging the slider until a 4 appears fits a fourth degree polynomial to the data in the scatterplot.

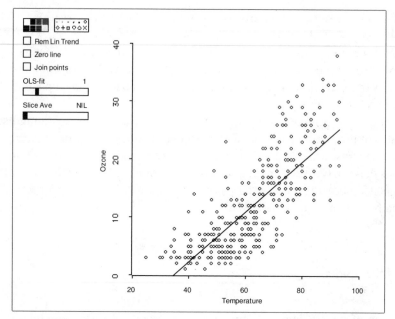

Figure 2.3. Plot of {*temperature, ozone*} with the ols fit superimposed.

2.3.2 Residual Plots

Another useful graphical method for checking linearity in simple regression
is to plot the residuals versus the predictor. This is obtained by clicking
on the button labelled "Rem Lin Trend," which stands for *remove linear
trend*. The vertical axis variable becomes a set of residuals, and the axis
label is changed to indicate that the residuals are from the ols regression
with *ozone* as the response and *temperature* as the predictor. The predictor
remains on the horizontal axis, as shown in Figure 2.4. The horizontal line
at zero on the vertical axis is often useful when interpreting residual plots.
It is obtained by pushing the button "Zero line" on the plot. For now ignore
the curved line superimposed on the plot; it is discussed in the next section.

We will use the notation $e|x$ to indicate the residuals given that x is fixed
at a selected value, just as we used the notation $y|x$ to indicate y given that
x is fixed at a selected value. If the simple linear regression model (2.2)
provides a good description of the data, then the distribution of $e|x$ should
not noticeably depend on x. In particular, if we were to examine individual
slices of a residual plot, then within each slice the mean should be close to
zero, and the variation should be constant.

Both the mean and the variance differ between slices in Figure 2.4,
so the linear regression model is not a good characterization of the ozone

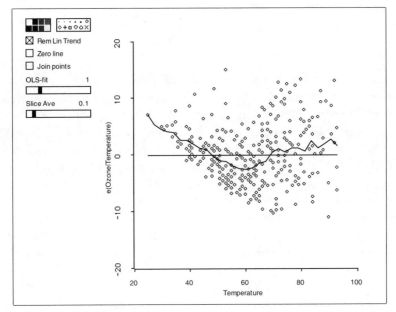

Figure 2.4. Residual plot for the ozone data with the zero line and an average smooth.

data. For example, consider slices around $x = 40$ and $x = 80$. A slice around $x = 40$ has mostly positive residuals and there is also a hint that the distribution of $e|(x = 40)$ may be positively skewed. Skewness in a residual plot often results when the distribution of y is bounded below and there are values near the bound. Since ozone level cannot be negative, the sharp lower left boundary of the plotted points in Figure 2.4 is expected. In contrast, a slice around $x = 80$ includes both positive and negative residuals, with much less evidence of skewness. Viewed as a whole, the plot in Figure 2.4 displays a characteristic U-shaped pattern that indicates nonlinearity. Nonconstant variance seems to be indicated as well.

2.3.3 Average Smoothing

The plots in the last section give convincing evidence that for the ozone data $E(y|x)$ is a nonlinear function of x, and the simple linear model (2.2) is not a good summary. These graphs do little to suggest just how $E(y|x)$ changes with x. Other graphical enhancements are needed to get a better feeling for $E(y|x)$ than is provided by the plot of the raw data in Figure 2.1.

Consider the slice about $x = 80$ in Figure 2.2, with window width of about 6. From the figure it appears that $E(y|x)$ is fairly constant within the slice and that any change is surely small relative to the within-slice standard deviation (SD) in ozone concentration, so

$$\frac{E[y|x = 83] - E[y|x = 77]}{\text{within-slice SD of } y}$$

is small. Without much loss of information about the mean, we can summarize the data in this slice by using the average of the responses, which is about 21 ppm, and the midpoint of the slice window, which is 80. Without the benefit of a model, 21 ppm is a useful estimate of $E(y|x = 80)$, the true mean ozone concentration at 80. Now take the summary point $(80, 21)$ and think of marking its position on a separate summary graph. Keeping the window width fixed, construct a new slice slightly to the left, centered at $x = 79$, summarize the data in the new slice as before, and again imagine plotting the summary point on the summary graph. The new summary point should be slightly below and to the left of the first point. This process can be repeated for every distinct value of x that occurs in the plot. The points on the summary graph will trace a curve that can give a good indication of just how $E(y|x)$ changes with the value of x. The curve may be easier to see after connecting adjacent summary points with lines. This process is called *smoothing*, and we will call the curve in the summary graph a *smooth* of the data. The curve resulting from application of the steps described here will be called an *average smooth* to reflect the fact that it is based on within-slice averages. A second type of smoothing will be described shortly.

The average smooth depends on the window width of the slices. Small window widths will produce undersmoothed jagged curves, while large window widths will produce oversmoothed curves that approach the horizontal line at the overall average response. A reasonable value for the window width that balances between undersmoothing and oversmoothing can usually be obtained by interactively changing the window width.

Average smooths are obtained in the *R-code* by using the slide bar labelled "Slice Ave," which is short for *slice average smoother*. Moving the slider to the right results in two actions. First, a standardized window width is displayed above the slide bar. The number shown is a fraction of the range of the data on the horizontal axis. In Figure 2.4, for example, the range of *temperature* on the horizontal axis is $93 - 25 = 68$. A standardized window width of 0.1 translates to an actual window width of $0.1 \times 68 = 6.8$. Second, a smooth corresponding to the displayed window width is superimposed on the plot. The curved and somewhat jagged line in Figure 2.4 is the smooth of the data corresponding to the smallest window width allowed by the *R-code*. Although it is rather rough, it still confirms the impression of a U-shaped trend in the residual plot. As the window width is increased, the jagged line in Figure 2.4 becomes smoother and gradually flattens, approaching the zero line for relatively large window widths.

2.3.4 Regression Smoothing

The average smoother uses the within-slice average to summarize the data in a slice, but other smoothers are possible as well. Instead of using the average response in a slice near $x = 80$, consider fitting a weighted regression using only a fraction f of the data with x-values closest to $x = 80$. For example, if $f = 0.25$, then the 25% of the data closest to $x = 80$ would be used. Points near $x = 80$ get higher weight than do distant points. The vertical coordinate of the first point in the summary graph is the fitted value for this local linear regression at $x = 80$. This is then repeated for every distinct value on the horizontal axis. This smoother is usually called *lowess* for *lo*cally *w*eighted *s*catterplot *s*moother.

To obtain the *lowess* smooth, hold down *both* the shift key and the mouse button while the cursor is on the "Slice Ave" slide bar. Select the item "lowess" from the resulting pop-up menu. Selecting this item will change the title of the slide bar to "lowess" and will change the smoother. The value displayed above the slide bar will now be f, the fraction of the data used to compute the fitted value at each observed value on the horizontal axis.

The menu obtained by shift-clicking on the smoother slide bar has five options: the slice average smoother, *lowess*, a weighted slice average smoother described in Section 2.4, an option to extract the smooth, and another option to fit a power curve. The last two options are described later in the book.

A *lowess* smooth for the ozone data is shown in Figure 2.5. The smooth captures the essential behavior of $E(y|x)$. It roughly consists of two linear phases with the change point around $x = 60$. The qualitative behavior of a *lowess* smooth is similar to that for an average smooth. Small values of f are likely to produce jagged undersmoothed curves, while relatively large values of f can produce oversmoothed curves that are close to the overall ols line.

Both average smooths and *lowess* smooths can help with the visual summarization of a scatterplot. They can be particularly helpful when there are many points in the plot, and one plotting location might correspond to many points; this is called *overplotting*. The smoothing process can be applied to a scatterplot of the raw data, as illustrated in Figure 2.5, or to a residual plot, as illustrated in Figure 2.4, or to any scatterplot where it may

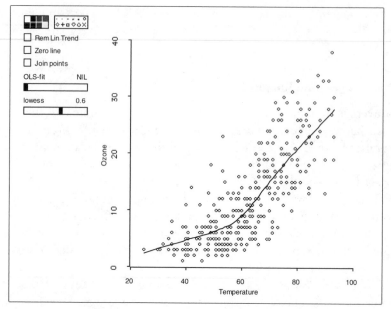

Figure 2.5. Plot of the ozone data with a *lowess* smooth.

be useful to characterize the regression function of the variable plotted on the vertical axis given the variable plotted on the horizontal axis.

2.4   COMPLEMENTS

Each of the plot controls introduced in this chapter acts on the points currently visible in the plot, without changing the underlying regression model. For example, pushing the "Rem Lin Trend" button will compute the ols regression of the vertical axis variable on the horizontal axis variable, using only the visible points, and then replace the vertical axis points by the residuals from this regression.

The smoothers described in Sections 2.3.3 and 2.3.4 are *kernel smoothers* and *regression smoothers*, respectively. Given a scatterplot $\{x, y\}$, a kernel smoother estimates a value y^* for a given value of $x = x^*$ as a weighted average, $y^* = \sum w_i x_i / \sum w_i$. Let $r_x = \max(x_i) - \min(x_i)$ be the range of the x_i in the data. For the average smoother, equal weight is assigned to all observations such that $|x_i - x^*|/r_x \leq d$ and weight zero to all other observations. The weighted average smoother, obtained by shift-clicking on the smoother slide bar and then selecting "Slice Wtd Average Smooth," assigns highest weight to an observation with $x_i = x^*$, and then linearly decreases the weight to zero for points with $|x_i - x^*|/r_x = d$. The value of the window width d is selected in the slide bar to be a number between zero and 1. There are many other kernel smoothers that one can use, some with better theoretical properties than the ones used here. Altman (1992) provides an introduction and Härdle (1990) a more technical discussion of kernel smoothers, including their use as an estimation method and the difficult technical problem of choosing an "optimal" window width. The idea of visually smoothing a scatterplot dates to at least Ezekiel (1930).

The *lowess* smoother was proposed by Cleveland (1979). The algorithm we use is given by step 1 of algorithm 6.1.1 in Härdle (1990, p. 192).

The ozone data have been discussed in Breiman and Friedman (1985). The air quality data used in Exercise 2.3 are taken from Chambers, Cleveland, Kleiner, and Tukey (1983). The data for Exercise 2.4 were provided by Ross Cunningham, Richard Telford and the Australian Institute of Sport.

EXERCISES

2.1. Consider adding a linear fit to Figure 2.4 by regressing the residuals on the predictor. Why will this result in nothing more than the zero line that is already on the plot?

2.2. Think of smoothing a scatterplot in which all the values of x are different. Describe the extreme behavior of an average smooth when the window width is chosen so small that each slice contains only one point. On every 2D plot produced with the *R-code* there is a button that can be used to superimpose this extreme average smooth on the plot. Which button is it?

2.3. The file `air.lsp` in the `R-data` folder contains data on several air quality measurements for 111 nearly consecutive days in the spring and summer of 1973 in the New York City area. Two of the variables in this data set are ozone concentration (*Ozone*) in parts per billion and the maximum daily temperature (*Temp*) in degrees Fahrenheit. The data on ozone concentration used earlier in this chapter were collected over a full year in the Los Angeles area in 1976.

Load the data file, and in the regression dialog specify *Ozone* as the response and *Temp* as the single predictor, and then push the "Done" button. From the resulting "Air" menu, use the "Plot of..." item to draw the plot {*Temp, Ozone*}. Use this graph to compare the relationship between temperature and ozone level for Los Angeles and New York City. Pay careful attention to the scales of the variables. Write a summary of your conclusions.

2.4. The file `ais.lsp` in the `R-data` folder contains data on 102 male and 100 female athletes collected at the Australian Institute of Sport. Load this data file, and when the regression dialog appears, specify height *Ht* as the single predictor and lean body mass *LBM* as the response, and then click the "Done" button. The resulting regression menu is called "BodyMass." The point labels for these data give the sex and sport of the athlete.

2.4.1. Using the "Plot of. . . " item from the "BodyMass" menu, construct the plot {*Ht*, *LBM*}, and identify the sport and sex of the tallest athlete. Find the sport and sex of the shortest athlete, and the two athletes with the highest *LBM*.

2.4.2. Using the plot {*Ht*, *LBM*} along with the plot controls discussed in this chapter, write a careful qualitative description of how the distribution of $LBM|Ht$ changes with the value of Ht. In particular, how do E(*LBM*|*Ht*) and var(*LBM*|*Ht*) change with the value of height? Does the simple linear regression model (2.1) appear to provide an adequate description? Provide support for your response by using the appropriate techniques of this chapter along with the corresponding plots.

CHAPTER 3

Two-Dimensional Plots

In this chapter, we discuss various plot controls and enhancements that can be used to understand 2D plots.

3.1 ASPECT RATIO AND FOCUSING

Not all scatterplots are equivalent for the purpose of gaining an understanding of the regression function, $E(y|x)$. Two scatterplots with the same statistical information can appear different because our ability to process and recognize patterns depends on how the data are displayed. At times the default display produced by a computer package may not be the most useful. One important parameter of a scatterplot that can greatly influence our ability to recognize patterns is the *aspect ratio*, the physical length of the vertical axis divided by that of the horizontal axis. Most computer packages produce plots with an aspect ratio near 1, but this is not always the best. The ability to change the aspect ratio interactively is important. We have already seen one example of this in Chapter 1.

As another example, consider Figure 3.1, which is a plot of the monthly U.S. births per thousand population for the years 1940–1948. The horizontal axis is labelled according to the year. The plot indicates that the U.S. birth rate was increasing between 1940 and 1943, decreasing between 1943 and 1946, rapidly increasing during 1946, and then decreasing again during 1947–1948. These trends seem to deliver an interesting history lesson

35

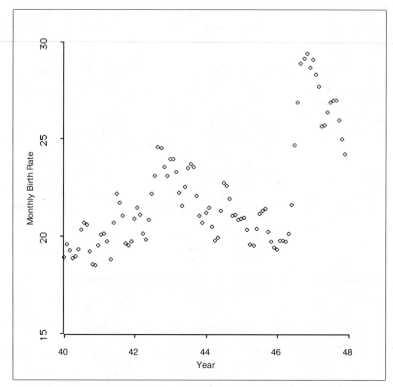

Figure 3.1. Monthly U.S. birth rate per 1000 population for the years 1940–1948.

since the U.S. involvement in World War II started in 1942 and troops began returning home during the first part of 1945, about nine months before the rapid increase in the birth rate. A duplicate of Figure 3.1 can be drawn with the *R-code* by selecting the "Birth Rates" item from the "Demo:2D" menu obtained by selecting `demo-2d.lsp` from the `R-data` folder. This plot is shown without the usual plot controls. They were removed by selecting the item "Plot Controls" from the plot's menu. Selecting this item again will restore the plot controls.

Let's now see what happens to Figure 3.1 when the aspect ratio is changed. Hold down the mouse button in the lower right corner and drag up and to the right. One reshaped plot is shown in Figure 3.2, which has an aspect ratio of about 1:4. The visual impact of the plot in Figure 3.2 is quite different than that in Figure 3.1. The global trends apparent in Figure 3.1 no longer dominate our visual impression. Also, Figure 3.2 displays many peaks and valleys. Is it possible that there are relatively minor within-year trends in addition to the global trends described in connection with Figure 3.1?

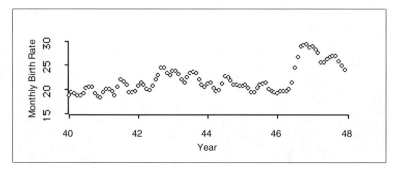

Figure 3.2. Monthly U.S. birth rate with a small aspect ratio.

To answer this question, we can *focus* on part of the data. Select the points corresponding to the years 1940–1943. Now from the plot's menu select the item "Focus on Selection." This will remove the points in the plot that are not currently selected. The menu item "Remove Selection" would, as the name implies, delete the selected points and leave the rest. Return to the "Plot" menu and select the item "Rescale Plot," which will recompute the values on the axes so the remaining data fill the plotting area. The result is shown in Figure 3.3. A within-year cycle is clearly apparent, with the lowest within-year birth rate at the beginning of summer and the highest occurring some time in the fall. This pattern can be enhanced by pushing the "Join points" button in the plot, causing adjacent points in the plot to be connected with a line. This gives the eye a path to follow when traversing the plot and can visually enhance the peaks and valleys.

The aspect ratio for the plot in Figure 3.3 is again about 1:4. To obtain this degree of resolution in a plot of all the data would require an aspect ratio of around 1:8. To return the plot of Figure 3.3 to its original state, choose the "Show All" item from the "Plot" menu and reshape it so that the aspect ratio is again about 1:1.

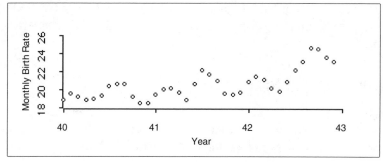

Figure 3.3. Monthly U.S. birth rate for 1940–1943.

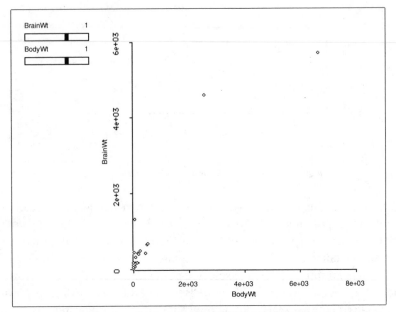

Figure 3.4. Plot of *BodyWt* versus *BrainWt* for 62 mammal species.

Changing the aspect ratio and focusing are useful methods for changing the visual impact of a plot, but they will not always work. Figure 3.4 contains a plot of body weight in kilograms and brain weight in grams {*BodyWt*, *BrainWt*} for 62 species of mammals. This plot can be obtained by selecting the item "Brain Weight Data" from the "Demo:2D" menu. The plot consists of three separated points and a large cluster of points at the lower left of the plot. It contains little visually available information about the dependence of the distribution of brain weight on body weight. Removing the three separated points and rescaling the plot helps a bit, but a large cluster near the origin remains. Repeating the procedure does not seem to help. The problem in this example is that the measurements range over several orders of magnitude. Body weight ranges from 0.01 to 6654 kg, for example. Transformations are needed to bring the data into usable form.

3.2   POWER TRANSFORMATIONS

The most common transformations are *power transformations*. A basic power transformation of a variable v is simply v^λ, where λ is the *transformation parameter*. For example, the notation $(BodyWt)^{0.5}$ refers to a variable with values equal to the square root of *BodyWt*. The transformation parameter can take any value, but the most useful values are nearly always found in the interval $-1 \leq \lambda \leq 2$. The variable v must be positive, or we will end up with complex numbers when λ is not an integer.

A slight variation of the basic power transformation is called a *scaled power transformation*, for which we use the notation $v^{(\lambda)}$. For a given λ, we define the transformation to be

$$
v^{(\lambda)} = \begin{cases} (v^\lambda - 1)/\lambda & \text{if } \lambda \neq 0 \\ \log(v) & \text{if } \lambda = 0 \end{cases} \tag{3.1}
$$

When $\lambda \neq 0$, the scaled power transformation differs from the basic power transformation only by subtracting 1 and dividing by λ. These changes have no important effect on the analysis, so when $\lambda \neq 0$, the scaled power transformations and the basic power transformations are practically equivalent. When $\lambda = 0$, the basic power transformation has the value $v^0 = 1$, but for the scaled power transformation $v^{(0)} = \log(v)$. Using this form therefore adds logarithms to the family of power transformations. Also, $v^{(\lambda)}$ is a continuous function of λ, so varying λ will not produce jumps when using the transformation sliders described next.

Return to the brain weight data displayed in Figure 3.4. The sliders on the plot can be used to choose a scaled power transformation. They are labelled according to the variable and the current value of the transformation parameter λ. Initially $\lambda = 1$, which means that no transformation has been applied. Each time you click on the slide bar the value of λ is changed and the plot is updated to display $v^{(\lambda)}$ in place of v, unless the new value of λ is 1. For this case, the original data are displayed. The labels on the axes are not changed after a transformation; to see the power, you must look at the numbers above the slide bars. Here and elsewhere in the *R-code*, λ can take the values ± 1, ± 0.67, ± 0.5, ± 0.33, 0, 1.25, 1.5, 1.75, and 2. These values of λ should be sufficient for most applications but they can be changed as described in Section A.3.

One possible transformed plot is shown in Figure 3.5, where the scaled power transformation with $\lambda = 0.33$ has been applied to *BodyWt* while the scaled power transformation with $\lambda = 0.5$ has been applied to *BrainWt*. Figure 3.5 is thus a plot of $\{BodyWt^{(0.33)}, BrainWt^{(0.5)}\}$. It doesn't give a very good representation of the data, although it does seem a little better than the plot of the untransformed data. Manipulate the sliders a bit and see if you can find a better pair of transformations for the brain weight data.

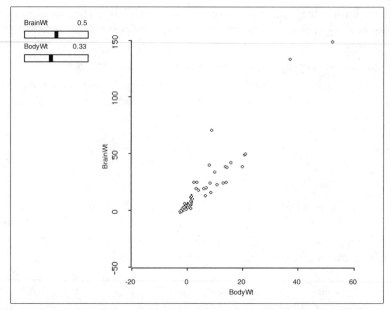

Figure 3.5. Both *BodyWt* and *BrainWt* are transformed according to the values given on the slide bar.

3.3   THINKING ABOUT POWER TRANSFORMATIONS

There are two simple rules that can make manipulating the power choice sliders easier:

- To spread the *small* values of a variable, make the power λ *smaller*.
- To spread the *large* values of a variable, make the power λ *larger*.

Most of the values of *BrainWt*$^{(0.5)}$ in Figure 3.5 are clustered close to zero, with a few larger values. To improve resolution, we need to spread the smaller values of *BrainWt*, so λ should be smaller. The values of *BodyWt*$^{(0.33)}$ are also mostly small, so we should spread the small values of body weight as well, again requiring a smaller power.

Decrease the transformation parameter for *BrainWt* in Figure 3.5 to $\lambda = 0.33$. The resulting plot is an improvement, but we still need to spread the small values of both *BodyWt* and *BrainWt*, and this indicates that we need to make both transformation parameters smaller still.

Set the transformation parameter for *BodyWt* to 0.5 and for *BrainWt* to -0.33, as shown in Figure 3.6. We now need to spread the larger values of *BrainWt*, and thus we should make λ larger. At the same time we need to spread the small values of *BodyWt*, and this is accomplished by decreasing λ. Figure 3.7 shows a plot of the brain data after applying the

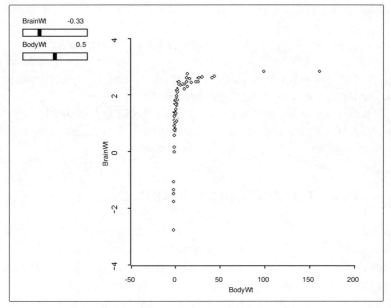

Figure 3.6. Another transformation of the brain weight data.

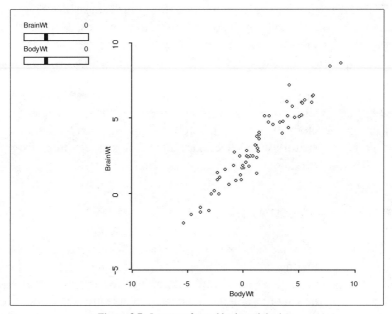

Figure 3.7. Log transformed brain weight data.

log transformation to both variables,

$$\{BodyWt^{(0)}, BrainWt^{(0)}\} = \{\log(BodyWt), \log(BrainWt)\}$$

This plot gives a good depiction of the data and strongly suggests that there is a linear relationship in the log-log scale.

Transformation sliders are not a regular part of the *R-code* plot controls for scatterplots. Instead, they are a part of the *R-code* plot controls for scatterplot matrices, discussed in the next chapter.

3.4   SHOWING LABELS AND COORDINATES

Selecting the "Show Labels" item from the plot's menu will cause point labels to be displayed on the plot when points are selected. When the "Show Labels" option is on, a check will appear beside the "Show Labels" menu item. To turn the option off, select "Show Labels" again. If no labels are supplied, the data points will be labelled according to the order in which they appear in the data file. *Xlisp-Stat* always starts numbering at zero, so the first case in the file is number 0, and the eleventh case is number 10. Labels are assigned to cases in the standard regression dialog in the *R-code*.

To find the coordinates of a point, select the item "Mouse Mode. . . " from the plot's menu. A dialog will appear with three choices. Choose "Show Coordinates" by pushing the corresponding button and then pushing "OK." The cursor will now change from an arrow to a hand with a pointing finger. Pointing at any data point and clicking the mouse button will show the point label and the coordinates as long as the mouse button is depressed; holding the Shift key while depressing the mouse button will make the coordinates remain on the plot. To remove the coordinates, shift-click again on the data point. To return to the selecting mode, choose "Selecting" from the "Mouse Mode" dialog.

3.5   LINKING PLOTS

Linking refers to tying two or more plots together, so that actions such as focusing, selecting, and deleting points in one plot are automatically applied in the others. The applicability of linking plots in simple regression problems is somewhat limited, but the birth rate data provide a convenient opportunity to introduce the idea.

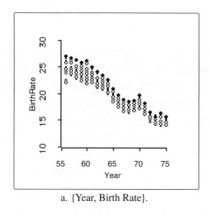
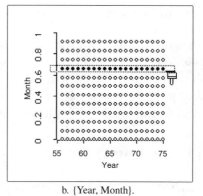

a. {Year, Birth Rate}.   b. {Year, Month}.

Figure 3.8. U.S. birth rate data.

We begin by constructing a plot of {*Year, BirthRate*} for the U.S. birth data between 1956 and 1975. The plot is shown in Figure 3.8a, which is constructed by plotting each data point against the year in which it was obtained, ignoring the month. This plot provides information on the distribution of monthly birth rates given the year but not the month. There is still one point for each month in the plot, but the correspondence between points and months is lost. Additional information might be obtained from Figure 3.8a if we could tell which points correspond to each month. There are a variety of ways to do this. One is to construct a linked plot of {*Year, Month*}, as shown in Figure 3.8b. This is a singularly uninteresting graph, but it is useful because it is linked to the plot in Figure 3.8a. Selecting all the points for any month in Figure 3.8b will cause the corresponding points in Figure 3.8a to be highlighted, as demonstrated for the month of September.

The plots in Figure 3.8 can be reproduced by selecting the "More Birth Rates" option from the "Demo:2D" menu. Choosing this menu item will start the *R-code* by displaying a regression dialog. Choose *BirthRate* as the response, and *Year* as the predictor; then push "Done." The variable *Names* is a list with entries like *Jun47* for the data point for June 1947. Since the items in this list are characters, the program guesses that these are to be used as case labels. Use the "Plot of. . . " item from the regression menu to construct the plots. These plots are automatically linked, as are most plots produced from a regression menu. To unlink a plot, select the "Unlink View" item from the plot's menu. To save space, the plots in Figure 3.8 are shown with the plot controls removed.

3.6   MARKING AND COLORING POINTS

The point symbol on a scatterplot can be changed using the *symbol palette*, and if your computer has a color monitor, the color can be changed using the *color palette*. First select the point or points you want to change, and then push the mouse button on a color or symbol in a palette. Colors and symbols are *inherited* by all plots linked to the plot you changed.

3.7   BRUSHING

An alternative method of selecting points is called *brushing*. In this method, the mouse pointer is changed into a selection rectangle. As the rectangle is moved across the screen, all points in the selection rectangle are highlighted, as are the corresponding points in all plots linked to it. As the rectangle moves, the selected points change.

To use brushing, select "Mouse Mode..." from the plot's menu, and select "Brushing" in the resulting dialog. After pushing the "OK" button, the mouse pointer is changed into a paintbrush, with an attached selection rectangle. The size and shape of the rectangle can be changed by selecting the "Resize Brush" item from the plot's menu and following the instructions given to resize the brush. Long narrow brushes are often the most useful.

In the plot shown in Figure 3.8b, change the mouse mode to brushing and make the brush a narrow horizontal rectangle. Then, brush across the plot from bottom to top. As you do so, examine the result in the plot shown in Figure 3.8a. The within-year trends are easy to spot as you move the brush. In particular, it appears that September always had the highest birth rate.

3.8   NAME LISTS

A *name list* is used to keep track of case labels. It is displayed as a separate window by selecting the "Show Case Names" item from the regression menu. Since this window is linked with all graphics windows, point selection, focusing, and coloring will be visible in this window.

3.9   COMPLEMENTS

Power transformations can be applied to a variable v that has negative values if a sufficiently large constant c is added before taking powers. The "Transformations..." dialog obtained from the regression dialog allows choosing a constant c in the text area marked "Center." The value entered must be large enough to make $v + c$ positive. The transformation produced is $(v + c)^\lambda$.

The definition of $v^{(0)} = \log(v)$ is justified by taking the limit as λ approaches zero.

Becker and Cleveland (1987) discuss scatterplot brushing and Stuetzle (1987) discusses plot linking. The birth rate data are taken from Velleman (1982). The brain weight data are from Weisberg (1985, p. 144).

The data that are obtained when loading the file `demo-2d.lsp` can also be obtained by loading a file from the `R-data` folder. The 1940–1947 birth rate data are in the file `birthrt1.lsp`, and the later birth rate data are in the file `birthrt2.lsp`. The brain weight data are in the file `brains.lsp`. The ozone example discussed in Chapter 2 is part of a much larger set of data in the file `ozone.lsp`.

EXERCISES

3.1. Take a closer look at the data of Figure 3.8 by using the graphical tools discussed in this chapter. Is it really true that the highest birth rate always occurred in September? Which month tends to have the lowest birth rates?

3.2. Redraw Figure 3.7. Which species has the highest brain weight and the highest body weight? Which species has the lowest brain weight and body weight? Find the point for humans on the plot and give the coordinates for the point.

3.3. This is a continuation of Exercise 2.4. Load the Australian Institute of Sport data, file `ais.lsp`, in the `R-data` folder.

3.3.1. Again specify *Ht* as the single predictor and lean body mass (*LBM*) as the response. It may be possible to explain more of the variation in *LBM* by using sex in addition to height. How does the distribution of $LBM|(Ht, Sex)$ change with the value of *Sex* and the value of *Ht?* You can gain insight into the answer by considering the plot {*Ht, LBM*} separately for males and females.

One of the variables in the data set is *Sex*, coded 1 for females and 0 for males. Using the "Plot of..." item in the regression menu, draw a histogram of the variable *Sex*. While this histogram is not particularly interesting on its own, it can be used to select points in the plot of {*Ht, LBM*} by sex. In the histogram, select all the points with the value of *Sex* equal to 1; this corresponds to selecting all the female athletes. The corresponding points are now selected in the scatterplot. Selecting the points with *Sex* equal to 0 selects only the male athletes. You can now use the "Focus on Selection," "Remove Selection," and "Show All" items in the plot's menu or in the histogram's menu to study the dependence of *LBM* on *Ht* for males and females separately.

As in Exercise 2.4 provide a report on your study, including the necessary graphical support.

3.3.2. Use the "New Model. . . " item to set up the regression structure with *%Bfat*, percent body fat, as the single predictor of *LBM*. Repeat Exercise 2.4 for these data. Investigate the distribution of *LBM* given *%Bfat*, separately for males and females, and summarize your results.

CHAPTER 4

Scatterplot Matrices

We now turn to graphical methods for regression problems with one or more predictors. In general, we have a response y and p predictors collected into a $p \times 1$ vector x. The individual predictors will be denoted by x_j for $j = 1, \ldots, p$, so $x = (x_1, \ldots, x_p)^T$. The general goal is unchanged: to study the conditional distribution of $y|x$ as the value of x changes, often concentrating on the regression function $E(y|x)$ and less frequently on the variance function $var(y|x)$. These functions now depend on p arguments, the values of the individual predictors. We have deliberately used the same symbol x to represent a *vector* of predictors in a p-predictor problem or a single predictor in simple regression. We will view regression with one predictor as the special case when $p = 1$.

When $p = 1$, the 2D plot $\{x, y\}$ provides a fairly complete summary of a regression problem. When $p = 2$, a 3D plot of the predictors versus the response can serve the same purpose. The 3D plot uses motion to view the third dimension, as described in the next few chapters. When $p > 2$, we cannot view the data in total because they are in too many dimensions. Consequently, graphical methods have been devised that allow the user to see parts of the data. One helpful graphical display is the *scatterplot matrix*, which provides an organized way of looking at many 2D views of higher dimensional data. This plot is a very useful starting point for regression graphics.

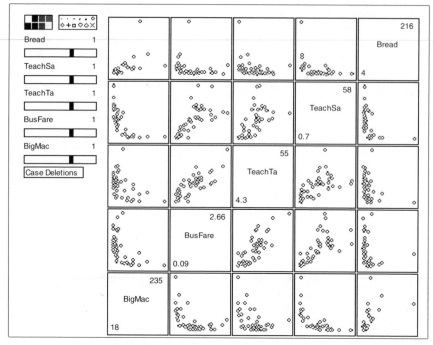

Figure 4.1. Scatterplot matrix of the Big Mac data.

4.1 USING A SCATTERPLOT MATRIX

A scatterplot matrix is a 2D array of 2D plots. To introduce the idea, load the file `big-mac.lsp` from the `R-data` folder. This file includes economic data on 45 world cities. For the purposes of this example, take as the response the variable *BigMac*, which is the number of minutes of labor required by an average worker to purchase a Big Mac hamburger and French fries. Use four predictors: *Bread*, the number of minutes of labor required to purchase one kilogram of bread; *TeachSal*, the average annual salary of a primary school teacher, in thousands of U.S. dollars; *TeachTax*, the tax rate paid by an average primary teacher; and *BusFare*, the lowest cost of a 10-kilometer bus, tram, or subway ticket, also in U.S. dollars. Using the regression dialog, select the response and predictors as specified above. The name of the regression will be "Mac" unless you change it.

Select "Scatterplot Matrix of. . . " from the "Mac" menu. In the resulting dialog, move variable names from the left window to the right in the order *BigMac*, *BusFare*, *TeachTax*, *TeachSal*, and then *Bread* by double clicking on the names; when done, push the "OK" button. The plot on your computer screen should look like Figure 4.1. You may wish to make the plot larger by resizing. You can't change the aspect ratio in scatterplot matrices.

Except for the diagonal, each frame of the matrix in Figure 4.1 contains a scatterplot. The variable names on the diagonal label the axes; the variables appear in the order selected beginning in the lower left and proceeding up the diagonal. The numbers in the diagonal cells are the maximum and minimum of the corresponding variable. *TeachTax*, for example, ranges between 4.3 and 55%, while *BusFare* varies from $0.09 to $2.66. The plots above the diagonal are mirror images of the plots below the diagonal. For example, the bottom right plot is {*Bread, BigMac*} while the top left plot is {*BigMac, Bread*}. Variable labels are truncated to fit along the diagonal.

A scatterplot matrix can be viewed as a graphical equivalent of a correlation or covariance matrix, as each plot shows the relationship between two variables without reference to the other variables. It can be a better summary of data than a correlation matrix since the latter gives only a single number summary of the *linear* relationship between variables, while each scatterplot gives a visual summary of linearity, nonlinearity, and separated points. For example, the plot of {*BusFare, TeachTax*} appears to be approximately linear, and so the relationship between these can be reasonably summarized by a correlation coefficient. The plot for {*TeachSal, TeachTax*} may also be approximately linear, but the variability appears to increase as *TeachTax* increases; this cannot be captured by a one-number summary. The 2D plots that include *Bread* appear to be nonlinear, although the impression of these plots is strongly influenced by at least one isolated point. The plots with *BigMac* on the vertical axis are curved.

4.2 IDENTIFYING POINTS

In several frames of the scatterplot matrix, one point appears to be separated from the others. Is it always the same point? To answer this, simply select the point and it will be highlighted in all frames of the scatterplot matrix. Assuming that you highlighted the point with the extremely high cost of bread, you will see that this point is very low on *TeachSal* and *TeachTax*, moderate on *BusFare*, and high, but not extreme, on *BigMac*. Which city is this? Select "Show Labels" from the plot's menu, and then select the point again: this city is Lagos, Nigeria. One might wonder if our impression of these plots would be altered if Lagos were removed from the plot. As with other graphs, a point can be removed by selecting first the point and then the menu item "Remove Selection" from the plot's menu. After selecting "Rescale Plot," the plot is shown in Figure 4.2. Removing

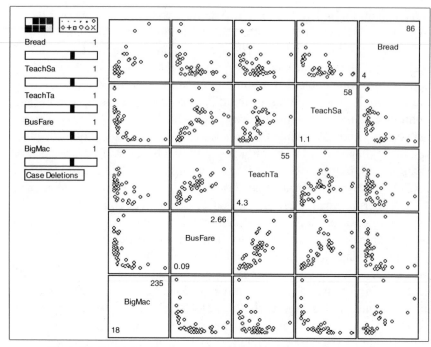

Figure 4.2. Rescaled scatterplot matrix of the Big Mac data after removing Lagos.

Lagos does not change most of the qualitative judgments concerning the bivariate relationships between these variables. Also, we see that the next largest value of *Bread*, for Manila, is 86 minutes.

We are now faced with a common problem in regression analysis: an apparently unusual point. A value of 216 for *Bread* means that the average worker must work for more than 3.5 hours to buy one loaf of bread. At this price, either bread must be a luxury item in the Nigerian diet or the value of 216 is in error, in either copying or computing. In any case, we choose to continue analysis without Lagos included in the data.

4.3 TRANSFORMING PREDICTORS TO LINEARITY

An important part of graphical methodology for regression is understanding the statistical relationships between the predictors. Nonlinear relationships between predictors make graphical methods difficult to interpret and use, an idea that will be developed in later chapters. Graphical methods are more effective when all regression functions of the form $E(x_j|x_k)$ for all j and k are linear functions of x_k, so all frames $\{x_k, x_j\}$ for the predictors in a scatterplot matrix show linear regression functions. Predictors can often be transformed to remove gross nonlinearities. The scatterplot matrix produced by the *R-code* can be used to help choose linearizing transformations for the predictors.

The main tools for choosing transformations of the predictors are the power transformation slide bars shown in Figure 4.2 and discussed in the last chapter. A slide bar is given for each strictly positive variable. Moving a slider replaces a variable in all frames of the scatterplot matrix by the scaled power transformation with the parameter given above the slide bar.

In Figure 4.2, each of the plots appears to be monotone, but not necessarily linear, so power transformations may be effective in achieving linearity. Consider first the plot {*TeachSal*, *Bread*}, near the upper right corner of the scatterplot matrix. For both variables, the majority of the observations are clustered near the axes. This suggests choosing powers that are small. Move the slide bars for *Bread* and *TeachSal* to 0.33. As you change the powers, every plot involving either of these variables is changed. The resulting plot {$(TeachSal)^{(.33)}$, $(Bread)^{(.33)}$} can now be judged for linearity. Because the plots in a scatterplot matrix are rather small, it is helpful to zoom in on this one frame by enlarging it in a standard 2D plot. Move the mouse over the plot you want enlarged, and *while holding down both the option and shift keys,*[1] push the mouse button. The computer will respond by beeping and then opening a new window containing an enlargement of the frame you chose. If the axes are transformed when you zoom in, they will be transformed in the 2D plot as well. If you option-shift-click on a variable name, a histogram of that variable or of its transformation will be opened in a window.

Figure 4.3 shows the enlarged 2D plot for *TeachSal* and *Bread* in the cube root scales. We can now use all the tools developed in the last two chapters to decide if this graph is linear or not. By removing the linear trend and then fitting a *lowess* smooth to the plot, it is apparent that this particular transformation does not quite achieve linearity. Try using logarithms rather than cube roots. The resulting plot is very nearly linear.

In the Big Mac data, linearity for *every* pair of predictors can be achieved, at least approximately, by replacing the predictors by their logarithms. This illustrates a general rule: Positive regression predictors that have the ratio between their largest and smallest values equal to at least 10 and preferably 100 or more should very likely be transformed to logarithms. In this example, *TeachTax* has maximum/minimum equal to about 10, and for the others this ratio is considerably more than 10, suggesting logarithms from the start.

---

[1]On Unix and Windows systems, hold down the Control and Shift keys.

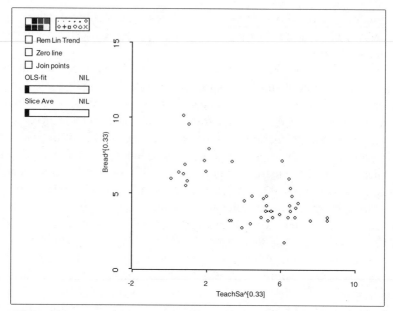

Figure 4.3. One frame from the scatterplot matrix for the Big Mac data without Lagos.

   The transformations of the predictors suggested here are designed to remove nonlinear relationships between pairs of the predictors. These have been chosen without reference to the response variable, and there is no guarantee that a transformation chosen to make the predictors linearly related will give a good scaling for the study of $y|x$. In any case, these linearizing transformations very often provide a good place to start an analysis.

4.4   PARTIAL RESPONSE PLOTS

Figure 4.4 gives the scatterplot matrix for the Big Mac data, still excluding Lagos, with all the predictors transformed to log scale. The last row of the figure gives the 2D plots of each of the transformed predictors versus the response. We call these *partial response plots* because they display the dependence of the response y on each predictor x_j without regard for any of the other predictors. Each of these four plots shows curvature to some degree. For example, the partial response plot {log(*TeachSal*), *BigMac*} suggests that *BigMac* decreases with log(*TeachSal*), but the decrease is not necessarily linear in log(*TeachSal*). To say more than this requires zooming in on this plot and examining it more closely; this is left to the homework problems. Similar statements can be made concerning the other

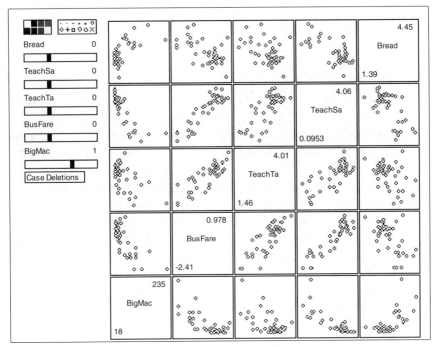

Figure 4.4. Scatterplot matrix for the Big Mac data with the predictors in log scale.

three partial response plots, and each of these can give useful information, particularly if the plot behaves in an unexpected way. For example, the cost of a Big Mac is lower in cities with high taxes, a conclusion that might not have been anticipated before looking at the plot.

The partial response plots display information about the partial regression of the response on each of the predictors. What can we learn from this row about the full p-dimensional regression function $E(y|x)$? The partial response plots always provide a visual lower bound for the goodness of fit that can be achieved with the full regression. If x_j does a good job explaining the response y, then a set of predictors that includes x_j can't do worse than x_j alone. Without further knowledge of the regression relationships among the predictors, the partial response plots tell only about bivariate relationships. For example, the partial response plots in Figure 4.4 are all curved, indicating that $E(y|x_j)$ is nonlinear for each predictor. This cannot necessarily be taken as evidence that the full regression function $E(y|x)$ is curved. Similarly, if each of partial response plots had been linear, we could not necessarily have concluded that the regression function $E(y|x)$ is linear. However, scatterplot matrices can be used to infer about the full regression function $E(y|x)$ when the predictors are linear, a theme that will be developed in later chapters (see Section 6.5).

4.5   COMPLEMENTS

Scatterplot matrices can also be used to display diagnostic statistics such as residuals and fitted values and other derived statistics like leverages and Cook's distances. For some of these statistics, primary interest is in the magnitude of the statistic, not in relationships with other diagnostics. The scatterplot matrix is then just a convenient way of displaying many plots at once. Adding the *Case numbers* variable to the plot can be useful.

Becker and Cleveland (1987) give further discussion of scatterplot matrices. The Big Mac data are taken from Enz (1991). The empirical rule to replace a strictly positive predictor by its logarithm if maximum/minimum exceeds some threshold like 10 or 100 has appeared in several places in the statistics literature, notably in Mosteller and Tukey (1977, p. 455). The data in Exercise 4.2 are taken from the 1992 *World Almanac* and *Information Please Almanac*.

EXERCISES

4.1.   *4.1.1.*   In the Big Mac data, find the cities with the most expensive Big Macs. Find the cities with the least expensive Big Macs. Where are bus fares relatively expensive? You might want to use a name list by selecting the "Show Case Names" item from the regression menu.

4.1.2.   By zooming in on some of the frames in the scatterplot matrix, verify that log transformations are effective in giving linear relationships between pairs of predictors. Would including Lagos in the plots have led to different conclusions concerning transformations of the predictors? Does Lagos still appear to be unusual with all predictors in the log scale?

4.2.   The data file `fuel90.lsp` in the `R-data` folder includes the data on fuel consumption in the 50 U.S. states in 1990–1991. Load this file. In the regression dialog choose as the response variable *FUEL/POP*, the per capita motor fuel consumption for the state. As predictors choose the number of vehicles per capita, *VEH/POP*; the tax rate on motor fuel per gallon, in cents, *TAX*; the miles travelled per vehicle, *VM/VEH*; and the per capita income in the state, *INC*.

Obtain a scatterplot matrix of the response and the four predictors. Are any unusual points apparent? For which states? Why are they unusual? Are transformations of the predictors needed to achieve linearity? How do you know? Are the partial response plots linear or nonlinear?

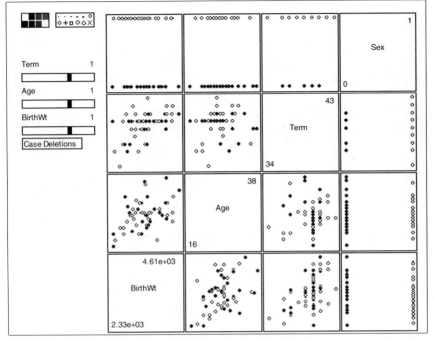

Figure 4.5. Scatterplot matrix of the Wellington birth weight data.

4.3. Data on 56 normal births at a Wellington, New Zealand, hospital are given in file `birthwt.lsp` in the `R-data` folder. The response variable for this problem is *BirthWt*, birth weight in grams. The three predictors are the mother's age, denoted *Age*; term in weeks, denoted by *Term*; and the baby's sex, denoted by *Sex*, 0 for girls and 1 for boys.

Load the file and then specify the response and the predictors in the regression dialog. Construct a scatterplot matrix of the variables *BirthWt*, *Age*, *Term*, and *Sex*, specifying the variables in the order given.

4.3.1. Describe the relationships in each of the partial response plots with emphasis on the individual regression functions. Why do the points in the scatterplots involving *Sex* fall in two lines?

4.3.2. Study the partial response plots {*Age*, *BirthWt*} and {*Term*, *BirthWt*} separately for each sex. This can be done by brushing the lines of points in any plot involving sex and observing the corresponding highlighted points in the other scatterplots, as illustrated in Figure 4.5. Is there visual evidence to suggest that the relationship between *BirthWt* and *Age* or *BirthWt* and *Term* depends on the sex of the baby? Describe the visual evidence that leads to your conclusion.

CHAPTER 5

Three-Dimensional Plots

In this chapter you will learn about the basics of 3D plotting. A 3D plot is specified by the notation $\{H, V, O\}$, which gives the names of the variables to be plotted on the three axes in the order {*H*orizontal axis, *V*ertical axis, *O*ut-of-page axis}, as shown in Figure 5.1. Three-dimensional plots are difficult to represent accurately on a 2D page, so using the computer while reading this chapter is particularly important.

5.1 VIEWING A THREE-DIMENSIONAL PLOT

Figure 5.2 is a 3D plot created with the *R-code*. It can be duplicated by loading the file `demo-3d.lsp` from the `R-data` directory and then selecting "View a surface" from the "Demos:3D" menu. The main difference between Figure 5.2 and the plot you produce is that the background and foreground colors are reversed: the motion of white points on a black background is easier to see than the usual black on white. If you want to change the background color, first select "Options. . ." from the plot's menu and then push the "White Background" button and finally push "OK." On the Macintosh and using Windows, the plot's menu is called "Spinner." The out-of-page dimension is not visible in Figure 5.2 because this dimension is perpendicular to the page of this book. To see the out-of-page dimension, it is necessary to rotate the plot.

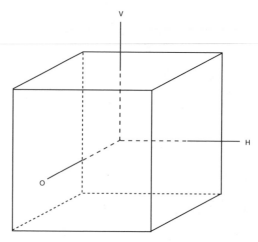

Figure 5.1. The 3D plotting region.

In the *R-code*, the plot's axes are always labelled with the letters H, V, and O. We will often use these letters as generic names for the variables on the axes as well. The names of the variables plotted on the axes are printed to the left of the plot; in Figure 5.2, these are X, Y, and Z.

5.1.1   Rotation Control

The basic tools for rotation control are three pairs of buttons at the bottom of the plot. Pushing any of these buttons will cause the points to appear to rotate. The two "Yaw" buttons cause the plot to rotate either to the left or right about the vertical axis of the computer screen. The V-axis of the plot may not always be the same as the stationary vertical axis of the computer screen. The "Roll" buttons cause rotation about the direction perpendicular to the computer screen, while "Pitch" rotates about the stationary horizontal axis of the computer screen. Holding down the shift key while pushing a control button causes the plot to rotate continually until one of the six control buttons is pushed again.

The rate of rotation is changed by selecting the "Faster" or "Slower" item in the plot's menu. These items can be selected more than once, and each selection will result in a slightly faster or slower rate of rotation.

Select the item "Mouse Mode. . ." from the plot's menu, and then select "Hand Rotate" in the resulting dialog. This changes the pointer into a hand; as you hold down the mouse button, you can use the hand to push the point cloud in various directions, much like you might push on the surface of a basketball to start it rotating. Pushing near the outside of the point cloud will result in relatively fast motion, while pushing near the center of the

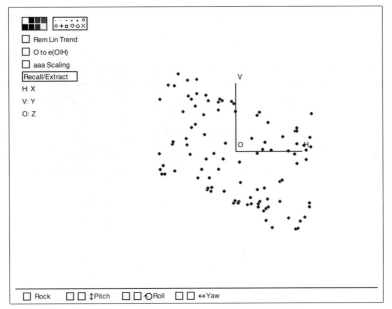

Figure 5.2. The initial view of the "View a surface" demonstration.

point cloud will result in relatively slow motion. Effective pushing takes some practice.

5.1.2 Recalling Views

The rectangular plot control marked "Recall/Extract" at the left of the plot is a button for a *pop-up menu*. Holding down the mouse button over "Recall/Extract" will produce a menu of options. The first item in this menu is "Recall Home," which will restore the plot to its original orientation. Selecting "Recall OLS" will rotate so the 2D plot in the plane of the computer screen is $\{\hat{V}, V\}$, where $\hat{V}$ denotes the fitted values from the ols regression of V on H and O, including an intercept term. The item "Remember view" will put a marker at any selected view. You can recall the view by selecting the item "Recall view." The other items in this menu will be described later in this chapter.

5.1.3 Rocking

During rotation a striking pattern may be seen, but it may be visible only while rotating. The impression of three dimensions can be maintained by stopping rotation near the viewing angle where the pattern is visible and then holding down the "Rock" button at the bottom left of the plot. As long as the mouse button is down, the plot will rock back and forth, allowing an impression of three dimensions to be maintained while staying near the interesting view.

Before rotation in Figure 5.2, the dominant feature is a linear trend; while rotating, distinct curvature emerges. These data were generated by setting X and Z each to be 100 uniform random numbers between $0°$ and $360°$ and computing $Y = \sin(X) + 2\cos(Z) + \sin(X)\cos(Z)$. From the rotating plot, one can clearly recognize shapes. Relating the shapes to a specific functional form is quite a different matter, however. Shapes will generally be less obvious in data analysis problems, but they can nonetheless be discovered.

5.1.4   Show Axes

Occasionally the axes in a 3D plot may be distracting. They are removed by selecting "Show Axes" from the plot's menu. Repeating this operation will restore the axes to the plot.

5.1.5   Depth Cuing

To create an appearance of depth, points in the back of the point cloud are plotted with a smaller symbol than are points in the front. As the points rotate, the symbol changes from small to large or vice versa. If some of the points are marked with a special symbol chosen from the symbol palette like an $\times$, then depth cuing is turned off. To turn depth cuing off by hand, select "Depth Cuing" from the plot's menu. Selecting this item again will turn depth cuing back on.

In all the 3D plots shown in this book depth cuing has been turned off.

5.2   SCALING AND CENTERING

The *plotting region* shown in Figure 5.1 is the interior of a cube centered at the origin, with sides running from -1 to 1. The data are *centered* and *scaled* to fit into this region. This can influence the interpretation of a 3D plot.

Suppose a quantity v is to be plotted on one of the axes. In the *R-code* and in most other computer programs, centering and scaling is based on the range $r_v = \max(v) - \min(v)$ and the midrange $m_v = [\max(v) + \min(v)]/2$. The quantity actually plotted is $2(v - m_v)/r_v$, which has minimum value -1 and maximum value 1. The centered and scaled variable fills the plotting region along the axis assigned to v. If the program is given an instruction to construct the plot $\{X, Y, Z\}$, what will really be produced is the plot

$$\{a(X - m_x), b(Y - m_y), c(Z - m_z)\} \tag{5.1}$$

where
$$a = 2/r_x, b = 2/r_y, \text{ and } c = 2/r_z$$

Centering usually has no effect on the interpretation of the plot. Scaling, however, can have an effect, and we refer to the operation leading to (5.1) as *abc-scaling*. Because of the scale factors a, b, and c in (5.1), we will not be able to assess relative size in plots produced with *abc*-scaling: a plot of $\{10X, 100Y, 1000Z\}$ will be identical to the plot $\{X, Y, Z\}$.

When relative size is important, it may be better to use *aaa-scaling*, in which the three scale factors in (5.1) are all replaced by the minimum scale factor, $\min(a, b, c)$. With *aaa*-scaling, the data on one of the axes will fill the plotting region, but the data on the other two axes may not. A plot of $\{10X, 100Y, 1000Z\}$ may appear quite different from the plot $\{X, Y, Z\}$ when *aaa*-scaling is used. When the plotted variables all have the same units, *aaa*-scaling of a plot can give additional useful information about the relative size and variation of the three plotted variables. In the *R-code*, *aaa*-scaling is obtained by pushing the "aaa Scaling" button. Repeating this operation will return to *abc*-scaling.

5.3 TWO-DIMENSIONAL PLOTS FROM THREE-DIMENSIONAL PLOTS

Use the "Recall/Extract" menu to return Figure 5.2 to the "Home" position and then rotate about the vertical axis by using the left "Yaw" button through roughly $60°$. The plot should now resemble Figure 5.3. The variable on the vertical screen axis is still Y, but the variable on the horizontal screen axis is some linear combination of X and Z. Which linear combination is it? After using the left "Yaw" button to rotate about the vertical axis through an angle of θ, the variable on the horizontal axis of the computer screen is

$$\begin{aligned} \text{horizontal screen variable} &= d + h \\ &= d + a(\cos\theta)X + c(\sin\theta)Z \end{aligned} \tag{5.2}$$

Here a and c are the scale factors used in (5.1), d is a constant that depends on the centering constants m_x and m_z and on the scale factors a and c, and h is the linear combination of the variables on the horizontal screen axis.

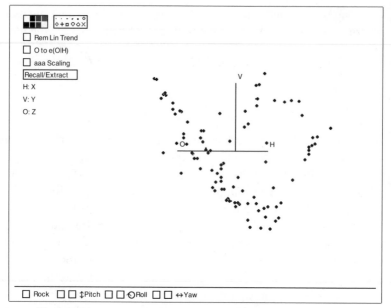

Figure 5.3. The same data is in Figure 5.2, rotated about the vertical by 60°.

Equation (5.2) might be understood by thinking of a circle in the horizontal plane. Each point on the circle determines a linear combination of X and Z that is h.

Since horizontal screen variables play an important role in statistical applications of 3D plots, the *R-code* allows you to print the values of d and h and also to save h as a variable for future calculations. Select the item "Print Screen Coordinates" from the "Recall/Extract" pop-up menu. This will print the linear combination of the quantities plotted on the horizontal and vertical screen axes. For the view in Figure 5.3, the printed output looks like this:

```
Linear combinations on screen axes in current rotating plot.
Horizontal: 0.378001  + 0.160653 H + 0 V + -0.28597 O
Vertical:   -0.00883192  + 0 H + 0.32905 V + 0 O
```

The H, V, and O refer to the original quantities (X, Y, Z in equation 5.1) plotted on the horizontal, vertical, and out-of-page axes. The quantity on the horizontal axis is $h \approx 0.16X - 0.29Z$ with the constant $d \approx 0.38$. Apart from a constant, the quantity on the vertical axis is proportional to Y. Your printed output may be slightly different because the rotation angle you use is not likely to be exactly 60°.

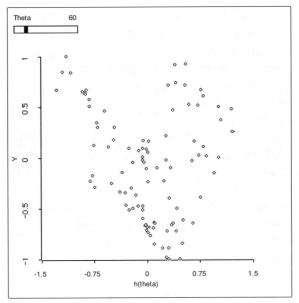

Figure 5.4. A 2D view of the "View a surface" demonstration. The slide bar for θ adds rotation to a 2D plot.

5.3.1   Saving h

The linear combination h on the horizontal screen axis can be saved for future calculations. Rotate a plot to have the horizontal screen axis that you want to save. Select the "Extract Horizontal" item from the "Recall/Extract" pop-up menu. You will be presented with a dialog to choose a name for this new quantity. The default name is h, and we will generally assume that you have used this default name. In practice the name should be changed to something meaningful since you may need to extract the horizontal screen variable several times in any particular problem. Push "OK" after you have typed a name. Once the horizontal screen variable is saved, further rotation of the plot will not change its value.

5.3.2   Rotation in Two Dimensions

The static view shown in Figure 5.3 is a *projection* of the points in the full 3D plot onto the plane formed by the vertical axis and the horizontal screen axis defined by (5.2). Since this is really a 2D plot, it could be viewed in a 2D scatterplot, as in Figure 5.4. We can use this figure to explain how rotation is done. When rotation is about the fixed vertical axis, we write

$$h = h(\theta) = a(\cos\theta)X + c(\sin\theta)Z$$

to recognize the dependence of h on the single angle θ of rotation. Imagine incrementing θ by a small amount; this will change $h(\theta)$ but will not change the variable plotted on the vertical axis. The computer screen can be refreshed by deleting the current points and redrawing the new points, $\{h(\theta), Y\}$. Repeating this process over and over gives the illusion of rotation. For a full 3D rotating plot, depth cuing and updated axes are all that need to be added.

Select the item "Rotation in 2D" from the "Demos:3D" menu to reproduce Figure 5.4, except that the rotation angle is initially set to $\theta = 0$. A slide bar marked "Theta" is added to this plot. As you hold down the mouse in this slide bar, θ and $h(\theta)$ are changed, and for each new θ the plot is redrawn. The slide bar changes θ in $10°$ increments, which is a much larger change than is used in the built-in rotating plot. As you change θ, the full 3D plot is visible. If you reverse the colors using the "Options..." item in the plot's menu, the rotation is a bit clearer.

We now see that 3D rotation about the vertical axis is nothing more than rapidly updating the 2D plot $\{h(\theta), Y\}$ as θ is incremented in small steps. Rotating a 3D plot once about the vertical axis by using one of the "Yaw" buttons corresponds to incrementing θ between $0°$ and $360°$. During rotation, plots of Y versus all possible linear combinations of X and Z are visible.

5.3.3   Extracting a Two-Dimensional Plot

Any 2D view of a 3D plot can be put into its own window by selecting the "Extract 2D Plot" item from the "Recall/Extract" menu. The new 2D plot has the usual plot controls, but it will not be updated when the parent 3D plot is changed.

5.3.4   Summary

The ideas of this section form a basis for viewing regression data in 3D plots, and so a brief summary is in order. While rotating a 3D plot $\{X, Y, Z\}$ once about the vertical axis, we will see 2D plots of *all* possible linear combinations of X and Z on the horizontal axis, with Y on the vertical axis. When the rotation is stopped, we see a 2D plot. The variable (5.2) on the horizontal screen axis can be extracted by using the "Extract Horizontal" item in the "Recall/Extract" menu. This variable will correspond to some particular linear combination of X and Z.

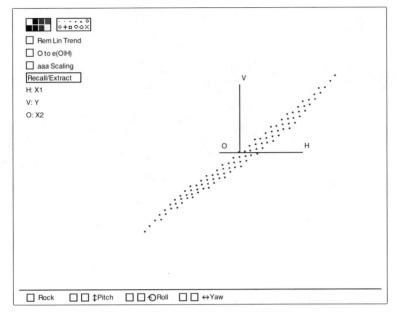

Figure 5.5. A 2D view of the "Detect a small nonlinearity" demonstration.

5.4 REMOVING A LINEAR TREND IN THREE-DIMENSIONAL PLOTS

A strong linear trend in a 3D plot may visually mask other interesting features, particularly nonlinearities. This problem can be overcome by removing the linear trend, leaving any nonlinear effects behind. In the *R-code*, a plot is *detrended* by pushing the "Rem Lin Trend" button on a 3D plot. This will replace the variable V on the vertical axis with $e(V|HO)$, the residuals from the ols regression of V on H, O, and an intercept. A detrended 3D plot is thus $\{H, e(V|HO), O\}$.

Figures 5.5 and 5.6 differ only by changing the vertical axis to give residuals rather than the original response. The deviations from the linear trend are evident in Figure 5.6 but nearly invisible in Figure 5.5, even while rotating. These data were generated by taking 100 points on a 10×10 grid for x_1 and x_2 and defining $y = u + \exp(-u)/(1 + \exp(-u))$, where $u = 0.909x_1 - 0.416x_2$ is a linear combination of the predictors. Figure 5.5 can be reproduced by selecting the "Detect a small nonlinearity" item from the "Demos:3D" menu. Figure 5.6 is obtained by pushing the "Rem Lin Trend" button. Pushing the button again will restore the plot to its original configuration.

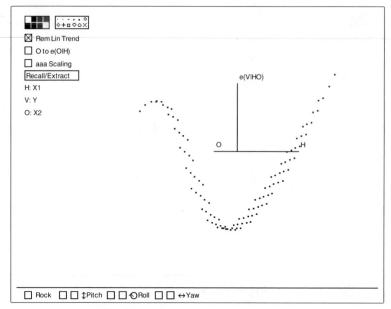

Figure 5.6. The same 2D view as in Figure 5.5, with the linear trend removed.

5.5   USING UNCORRELATED PREDICTORS

Select the item "Colinearity hiding a curve" from the "Demos:3D" menu and rotate the resulting 3D plot. What do you observe? The points in the plot appear to fall close to a rotating vertical sheet of paper. One static view of this plot is given in Figure 5.7. This figure and the full rotating plot are of little help in finding structure because the predictors x_1 and x_2, plotted on the H and O axes, respectively, are highly correlated. To see this clearly, return the plot to the "Home" position and then use the "Pitch" control to rotate to the 2D plot $\{H, O\}$. The fact that the predictors are highly correlated should now be apparent, as all the points fall very close to a line. Finding structure in 3D plots is likely to be difficult when the predictors on the H and O axes are even moderately correlated.

With highly correlated predictors, some 2D views of the 3D plot lack sufficient resolution for any structure to be clearly visible. This is the case in Figure 5.7, where the values of the horizontal screen variable are tightly clustered about the origin of the plot. To insure good resolution in a 3D plot, we would like the values of the horizontal screen variable to be well spread in every 2D view. This can be accomplished by replacing the current predictors with an equivalent pair of *uncorrelated predictors*. The resulting new plot will be of the generic form $\{H, V, O_{new}\}$, so we will

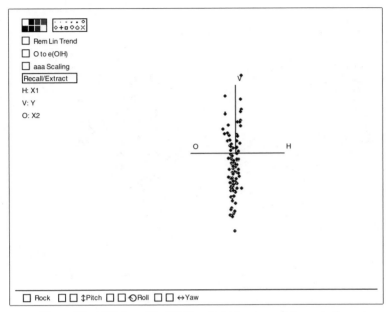

Figure 5.7. A 2D view of the "Collinearity hiding a curve" demonstration.

need to change only the variable on the out-of-page axis. The new variable O_{new} is the residuals from the ols simple linear regression of O on H, including an intercept, $e(O|H)$. Since the sample correlation between the residuals $O_{new} = e(O|H)$ and H is zero (recall Exercise 2.1), the variables on the horizontal and out-of-page axes of the new plot will be uncorrelated, and the values of the horizontal screen variable should be well spread in any 2D view.

Beginning with any 3D plot, a new plot with uncorrelated predictors is obtained by pushing the "O to e(O|H)" button. Pushing the button again will restore the plot to its original state.

Changing to uncorrelated predictors in Figure 5.7 results in Figure 5.8, where a curved trend is now visible. In this example, the predictors have correlation close to 0.99, and the response y is a function of x_1, x_2 and x_1^2, x_2^2 and x_1x_2.

We gain resolution when changing to uncorrelated predictors in a 3D plot, but do we lose information? Since the residuals $e(O|H) = O_{new}$ from the regression of O on H are computed as a linear combination of O and H, rotating the new plot $\{H, V, e(O|H)\}$ will still display 2D plots of all possible linear combinations of O and H, just as we obtain when rotating the original plot $\{H, V, O\}$. No graphical information is lost when changing to uncorrelated predictors.

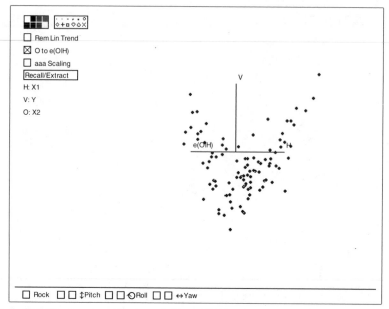

Figure 5.8. The same demonstration as in Figure 5.7, but with the "O to e(O|H)" button pushed.

5.6   COMPLEMENTS

Plot rotation was introduced in the statistics literature by Fisherkeller, Friedman, and Tukey (1974). Many early papers on this subject are reprinted in Cleveland and McGill (1988). Tierney (1990, Section 9.1.3) provides a useful reference for many of the details of plot rotation. The need for uncorrelated predictors in a rotating plot was presented by Cook and Weisberg (1989, 1990a). In the latter paper, an optimality property of this procedure is derived. The discussion of rotation in Section 5.3 is limited to rotation about the vertical axis, but it is easily generalized.

The RANDU generator discussed in Exercise 5.2 is taken from Tierney (1990, p. 41). Exercise 5.3 is based on a demonstration included with *Xlisp-Stat*, also by Luke Tierney. The haystack data in Exercise 5.4 are from Ezekiel (1941). The data for Exercise 5.5 are taken from Tuddenham and Snyder (1954).

EXERCISES

5.1.   In Figure 5.4, the tick-marks on the horizontal axis cover the range from −1.5 to 1.5, even though each of the variables displayed in the plot are scaled to have values between −1 and 1. Why is this larger range necessary?

5.2. Many statistical methods use a sequence of numbers that behave as if they were a random sample from some specified distribution. The most important case is generating a sample from the uniform distribution, so each draw from the distribution is equally likely to be any number between 0 and 1. The random-number generators used are deterministic, so if you start the generator in the same place, it will always give the same sequence of numbers. Consequently, generators must be tested to see if the deterministic sequences they produce behave as if they are a random sample from a distribution.

A well-known generator of uniform random numbers in the early days of computing is called RANDU. Load the file `randu.lsp` from the `R-data` folder. This will give you a 3D plot obtained by taking 200 consecutive draws (x_1, x_2, x_3) from RANDU and then plotting $\{x_1, x_2, x_3\}$.

5.2.1. As you spin the plot, the points fall in a cube. Is this expected, or is it evidence that RANDU is not generating numbers that behave like a uniform sample?

5.2.2. If the numbers all behaved as if they were draws from a uniform distribution, what would you expect to see in every 2D view of this plot? Spin the plot, and see if you can find a 2D view that is not random. You may need to spin slowly to find anything. What do you conclude?

5.3. Load the file `spheres.lsp` from the `R-data` folder. This will automatically produce two 3D plots. Move one so both are visible at the same time. Start both rotating by pushing the mouse button in the left "Yaw" button while pressing down the Shift key. What is the difference between these plots? If you are having trouble finding any difference, select a vertical strip in each plot by brushing over the strip with the mouse button down, and then examine the strips as the plots rotate.

5.4. In the Great Plains during the 1920s, farmers sold hay unbailed and in the stack, requiring estimation of the volume of a stack to insure a fair price. Two measurements that could be made easily with a rope were usually employed: *Circum*, the circumference in feet around the base of the stack, and *Over*, the distance in feet from the ground on one side of a stack to the ground on the other side of the stack. Stacks varied in height and shape so using a simple computation like the volume of a hemisphere, while perhaps a useful first approximation, was not found to be sufficiently accurate.

The file `haystack.lsp` in the `R-data` folder contains measurements on *Circum* and *Over* for 120 haystacks in Nebraska in 1927 and 1928. The response variable is *Vol*, the true volume of the stack obtained by using several careful measurements. The measurement process for obtaining *Vol* was not available for the average Nebraska farmer at the time.

Load the haystack file, and when the regression dialog appears, specify *Vol* as the response and *Circum* and *Over* as predictors. From the resulting "Haystacks" menu use the item "Plot of..." to construct the 3D plot {*Circum*, *Vol*, *Over*}: Move the variables *Circum*, *Vol*, and *Over* in this order from the left "Candidates" window to the "Selected Axes" window by double clicking on the variable names. If you make a mistake, double clicking moves the variables back as well. Push "OK" when you are finished.

5.4.1. Inspect the 3D plot {*Circum*, *Vol*, *Over*} by using only the rotation controls along the bottom of the plot. Write a two- or three-sentence description of what you see as the main features of the plot.

5.4.2. Change to *aaa*-scaling in the 3D plot {*Circum*, *Vol*, *Over*}. Write a brief description of the result and why it might have been expected based on the nature of the measurements involved.

5.4.3. Remove the linear trend from the plot {*Circum*, *Vol*, *Over*} by clicking the "Rem Lin Trend" button. Describe the quantity on the vertical axis of the detrended plot. Write a two- or three-sentence description of the main features of the detrended plot.

5.5. This problem uses data from a study of the growth of children born in Berkeley, California, in 1928–1929. We will use only data on girls. Load the file `BGSgirls.lsp` from the `R-data` folder, and select the age 18 weight, *WT18*, as the response and the age 2 weight, *WT2*, and the age 9 weight, *WT9*, as the predictors. All measurements are in kilograms.

5.5.1. Use the "Plot of..." item in the regression menu to draw the plot of {*WT2*, *WT18*, *WT9*}. Inspect this plot using only the rotation controls along the bottom of the plot. Write a two- or three-sentence description of what you see.

5.5.2. Change to *aaa*-scaling in this 3D plot. Explain what happened. Does the scaling convey additional information? What is the information? Remember that all three variables are measured in the same scale.

5.6. Return to the `birthwt.lsp` data described in Problem 4.3 and construct the 3D plot {*Age*, *BirthWt*, *Term*}. Inspect the plot using any of the plot controls discussed in this chapter and write a brief description of your impressions.

CHAPTER 6

Visualizing Linear Regression with Two Predictors

In regression problems with one predictor we study how the distribution of $y|x$ changes as the value of x changes. We consider two predictors in this chapter, studying how the distribution of $y|(x_1, x_2)$ changes with the values of x_1 and x_2. As in the case of a single predictor, the primary emphasis will be on the behavior of the regression function $E(y|x_1, x_2)$. The basic graphical construction for this study is a rotating 3D scatterplot, $\{x_1, y, x_2\}$. We first review the linear regression model for two predictors, which is assumed to hold throughout this chapter.

6.1 LINEAR REGRESSION

The linear regression model with two predictors is an extension of the model for a single predictor discussed in Section 2.2,

$$y|(x_1, x_2) = \beta_0 + \beta_1 x_1 + \beta_2 x_2 + \varepsilon \tag{6.1}$$

The β's are unknown regression coefficients. The errors ε are assumed to be independent of x and of each other with mean zero and constant variance σ^2. In vector notation, the model is

$$y|x = \beta_0 + \beta^T x + \varepsilon \tag{6.2}$$

where we have collected $x = (x_1, x_2)^T$ into a vector of length 2 and $\beta^T = (\beta_1, \beta_2)$. In the notation we are using here the vector of unknown regression

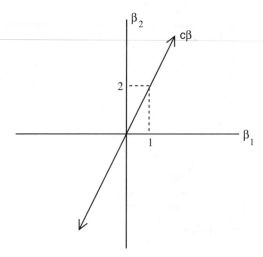

Figure 6.1. In linear regression with two predictors, β may be any point in the plane. When $c\beta^T = (1, 2)$, β is constrained to lie along the line shown in the figure.

coefficients β does not contain the intercept β_0, so x and β have the same length.

There are two key implications of the linear model. First, the distribution of $y|x$ depends on x only through the linear combination $\beta^T x$. As a result, the distribution of $y|x$ is the same as the distribution of $y|\beta^T x$ for all values of x, and the regression functions for $y|x$ and $y|\beta^T x$ must be identical, $E(y|x) = E(y|\beta^T x)$. The second implication is that these regression functions are linear:

$$\begin{aligned} E(y|x) &= \beta_0 + \beta_1 x_1 + \beta_2 x_2 \\ &= \beta_0 + \beta^T x \end{aligned}$$

In this equation, β may be any point in the plane in Figure 6.1.

6.1.1   The Ideal Summary Plot

Suppose we know the product $c\beta$ for some unknown constant $c \neq 0$. This supposition is relevant because we will eventually be able to estimate $c\beta$ graphically, but not β itself. Knowing $c\beta$ restricts the possible values of β. For example, if $c\beta^T = (1, 2)$, then the value of β must lie on the line shown in Figure 6.1. We can now compute a new predictor $h^* = c\beta^T x$ for each value of x in the data, and rewrite $E(y|x) = E(y|\beta^T x)$ as

$$E(y|h^*) = \beta_0 + c^{-1} h^* \tag{6.3}$$

Knowing $c\beta$ has allowed us to reduce the original regression to a simple linear regression problem without any loss of information.

The data for regression model (6.3) can be visualized in the 2D plot $\{h^*, y\}$ just as we visualize the data for any regression problem with one predictor. We call the plot $\{h^*, y\}$ an *ideal summary plot* because it provides full information on the original regression problem with two predictors. Model (6.3) is a consequence of assuming the linear regression model, but summarizing a regression problem with a 2D plot is a general idea that does not require a model. In this chapter we study 2D summary plots when a linear model holds, allowing us to introduce several new ideas in a familiar setting.

6.1.2   Viewing an Ideal Summary Plot When $\sigma^2 = 0$

Given a linear model like (6.1), imagine how the rotating 3D plot $\{x_1, E(y|x), x_2\}$ would look. In real problems we can never construct this plot because $E(y|x)$ is unknown, but to fix ideas, we can generate data from such a model and then examine it. We generated a data set with 100 observations according to the following setup. The predictors x_1 and x_2 were first generated as independent N(0, 1) random variables, where the notation $N(\mu, \sigma^2)$ denotes a normal random variable with mean μ and variance σ^2. The response was then constructed as

$$y|x = 1 + 2x_1 + 3x_2 + N(0, 1)$$

The errors are independent, normal random variables with mean 0 and variance 1. The regression function is thus $E(y|x) = 1 + 2x_1 + 3x_2$ with $\beta = (2, 3)^T$. The data set can be obtained by loading the file demo-3d.lsp from the R-data folder and then selecting the item "Uncorrelated Predictors, Linear Model" from the menu. You will be presented with a regression dialog containing the two predictors, y, $E(y|x)$ and the errors. Specify a regression with response y and predictors x_1 and x_2. The name of the resulting regression menu will be "Uncorr," unless you change it.

Use the "Plot of. . ." item in the "Uncorr" regression menu to draw the 3D plot $\{x_1, E(y|x), x_2\}$. Inspect the rotating plot for a bit by using the "Yaw" plot controls to rotate around the vertical axis. What do you see as the important characteristics of the plot? Most striking is that all the points fall exactly on a plane that is visible while rotating. The plane is 2D because there are two predictors; it is p dimensional when the number of predictors is p.

Figure 6.2 shows a particular 2D view with all the points falling exactly on a line. For this plot the horizontal screen variable $h^* = c\beta^T x = c(2x_1 +$

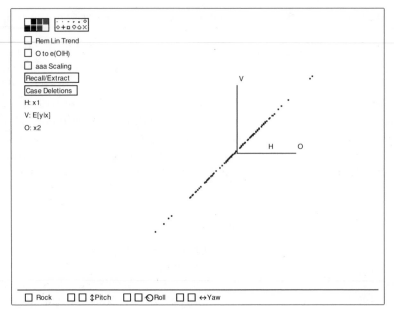

Figure 6.2. A 2D view of the "Uncorrelated Predictors, Linear Model" demonstration, with $E(y|x)$ on the vertical axis.

$3x_2$). Because the linear model holds, the constant c is determined by the scaling and centering factors used to map the data into the plotting area. This view is an ideal summary plot.

When rotation is stopped and the horizontal screen variable h is *not* proportional to $\beta^T x$, the points will not fall exactly on a line. This brings us to another important characteristic of the example: Each static 2D view shows linearity between the response and the horizontal screen variable. The strength and slope of the linear relationships vary, but no 2D view shows curvature. This is a consequence of both the linear model and the way we selected the predictors. Curvature can be obtained for other distributions of the predictors, as will be described later in this chapter.

6.2   FITTING BY EYE

6.2.1   Fitting by Eye When $\sigma^2 = 0$

Return to the 3D plot $\{x_1, E(y|x), x_2\}$ and rotate it until the static 2D plot in the computer screen gives a single straight line, as in Figure 6.2. The view can be fine tuned by selecting the item "Slower" from the plot's menu to slow the rate of rotation. You have just fitted a linear regression model by eye; the answer is the 2D plot in Figure 6.2. Fitting by eye is easy in this case because there is no error.

The new predictor is the horizontal screen variable of Figure 6.2, apart from the unimportant additive constant. This linear combination of the predictors can be determined by using the "Print Screen Coordinates" item in the "Recall/Extract" pop-up menu:

```
Linear Combinations on screen axes in current rotating plot.
Horizontal: -0.043509  + 0.199105 H + 0 V + 0.298658 O
Vertical:   -0.0940762  + 0 H + 0.096433 V + 0 O
```

Your output will be different if you stopped rotation at a different view. We get

$$h^* = b_1 x_1 + b_2 x_2 = 0.199105 x_1 + 0.298658 x_2$$

as our single new predictor. Apart from rounding error, all points in Figure 6.2 fall exactly on a line and thus equation (6.3) is satisfied. Another check is to compare β_1/β_2 to b_1/b_2; these ratios will be equal if $b = c\beta$. In this example, $b_1/b_2 = 0.199105/0.298658 = 0.67$, and $\beta_1/\beta_2 = 2/3 = .67$.

We can use equation (6.3) to recover the exact form of $E(y|x)$ as a linear function of x_1 and x_2 by performing the simple linear regression of $E(y|x)$ on h^*. This regression can be carried out in the *R-code* as follows. Make sure that the 3D plot $\{x_1, E(y|x), x_2\}$ is still rotated to display a straight line on the computer screen and then select the item "Extract Horizontal" from the "Recall/Extract" pop-up menu. Enter the name *h-star* in the dialog and then push "OK." You have just saved *h-star* as part of the data file. To carry out the regression of $E(y|x)$ on *h-star*, return to the "Uncorr" menu and select "New Model. . . ." Specify *h-star* as the only predictor and $E(y|x)$ as the response and then push "Done." The output is shown in Table 6.1.

The standard errors and $\hat{\sigma}$ are not quite zero in this output because of rounding errors in the computer; if you chose a different linear combination, your estimates and standard errors will be different. The ols intercept and slope are 1 and 10.0449, respectively, and

$$E(y|x) = 1 + 10.0449 h^*$$

From (6.3), we compute $c^{-1} = 10.0449$ and thus $c = 0.0995035$. Finally, replacing h^* with its representation in terms of the predictors, we recover the true regression function:

$$\begin{aligned} E(y|x) &= 1 + 10.0449(0.199105 x_1 + 0.298658 x_2) \\ &= 1 + 2x_1 + 3x_2 \end{aligned}$$

Table 6.1. Regression of y on h^*

```
Model name = Uncorr1, Response = E[y|x]
Coefficient Estimates
Label                    Estimate    Std. Error       t-value
Constant                        1   4.82313e-09   2.07334e+08
h-star                    10.0449   1.26497e-08   7.94084e+08

R Squared:                      1
Sigma hat:             4.81678e-08
Number of cases:              100
Degrees of freedom:            98
```

In practice we will rarely know $c\beta$ so we cannot usually draw an ideal summary plot. But we can graphically determine an estimate h of h^* using 3D plots.

6.2.2   Fitting by Eye When $\sigma^2 > 0$

Return to the "Uncorr" menu and use the "Plot of..." item to get the 3D plot $\{x_1, y, x_2\}$. How does this plot differ from the plot $\{x_1, E(y|x), x_2\}$? The points in the 3D plot $\{x_1, y, x_2\}$ scatter about a plane, as is evident during rotation. They do not fall exactly on a plane because of the errors. Rotate the plot using a "Yaw" button to get a good feeling for the point cloud, and then stop rotation. The resulting static 2D view shows a linear relationship. Some static views show strong linearity and some show only weak linearity. Rotate the 3D plot to the 2D view that shows the strongest linear trend. Your plot should look like Figure 6.3 or its mirror image. Select the item "Remember view" from the "Recall/Extract" pop-up menu; this will allow you to return to this view after further rotation by selecting the item "Recall view." The linear combination of the predictors that defines the horizontal axis of the 2D plot should be close to $h^* = c(2x_1 + 3x_2)$ for some constant c, and the ratio of the coefficients should be close to 2/3. Because there is error, we should not expect these results to be exact, as they were in the no-error case discussed previously. Selecting "Print Screen Coordinates" from the "Recall/Extract" pop-up menu, we get

```
Linear Combinations on screen axes in current rotating plot.
Horizontal: -0.0418409  + 0.204554 H + 0 V + 0.296553 O
Vertical:   -0.0335182  + 0 H + 0.0871291 V + 0 O
```

The value of the horizontal screen variable h is almost equal to $0.1 \times (2x_1 + 3x_2)$, and the ratio

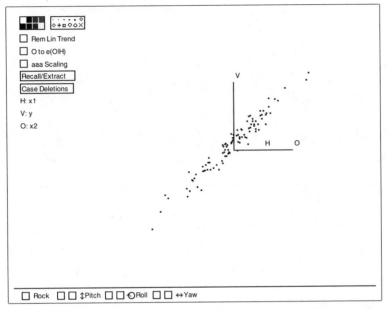

Figure 6.3. Strongest 2D view from the "Uncorrelated Predictors, Linear Model" demonstration, with y on the vertical axis.

$$0.204554/0.296553 = 0.689772$$

is not far from 2/3. We have nearly recovered the true linear combination of the predictors that determines the response. In problems with larger errors, graphically choosing a best direction can become more difficult; ols fitting becomes more variable as well.

6.2.3 Fitting by ols

The ols estimates $\hat{\beta}$ and $\hat{\beta}_0$ are obtained by minimizing the sum of squared differences between the observed values of the response y and the corresponding fitted values. The ols plane includes all the points $\{x_1, \hat{y}, x_2\}$, where $\hat{y} = \hat{\beta}_0 + \hat{\beta}_1 x_1 + \hat{\beta}_2 x_2$ are the fitted values and $\hat{\beta}_0, \hat{\beta}_1, \hat{\beta}_2$ are the ols estimates. Thus fitting by ols will determine a linear combination $h_{\text{ols}} = \hat{\beta}_1 x_1 + \hat{\beta}_2 x_2$. Imagine summarizing the 3D plot by using the 2D plot $\{h_{\text{ols}}, y\}$, or equivalently, $\{\hat{y}, y\}$. What would this plot look like? If $\hat{\beta}$ is a good estimate of β, then equation (6.3) should be nearly satisfied, with h_{ols} substituted for h^*.

Select the item "Recall OLS" from the "Recall/Extract" pop-up menu in the plot $\{x_1, y, x_2\}$. The view on the computer screen is now $\{h_{\text{ols}}, y\}$. By first selecting "Recall view" and then "Recall OLS," you can compare

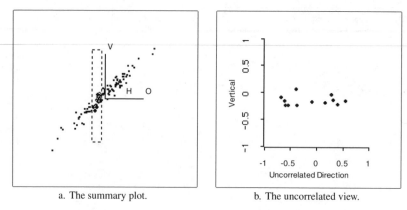

a. The summary plot.	b. The uncorrelated view.

Figure 6.4. Assessing the summary plot for Figure 6.3.

the view $\{h, y\}$ you saved earlier in Section 6.2.2 with the ols view. If you did a good job at choosing the strongest linear trend by eye, the ols view and the view you saved in Section 6.2.2 should be nearly the same, or they may be mirror images of each other.

We can substitute visual fitting in a 3D plot for ols. As we rotate the 3D plot about the vertical axis, we see a sequence of 2D plots $\{h, y\}$, where $h = b_1 x_1 + b_2 x_2 = b^T x$ for some vector b. We hope to find $b \approx c\beta$ for some nonzero constant c, stopping rotation when the strongest linear trend is visible on the computer screen. If we do find $b \approx c\beta$, then (6.3) will be approximately satisfied with h substituted for h^*.

6.2.4   Checking a Candidate Summary Plot

As long as the linear model holds, an ideal summary plot $\{c\beta^T x, y\}$ can replace the full 3D plot without loss of any statistical information. A *summary plot* $\{h, y\}$ is an estimate of an ideal summary plot, with h chosen using the methods described in the last two sections. Since h is determined using data, we need a method of checking to see if a particular *candidate summary plot* $\{h, y\}$ is adequate.

If h is a good approximation to $h^* = c\beta^T x$, then the 2D summary plot contains essentially all the information on how the distribution of $y|x$ changes with the value of x. This is the same as saying that the distribution of $y|h$ is independent of x, which provides the basis for checking to see if a candidate summary plot misses important information. Imagine selecting points in a vertical slice of $\{h, y\}$. The summary plot shown in Figure 6.3 is repeated in Figure 6.4a, with a slice selected. If h is all we need to know about the predictors, then the selected points within the slice should look like a horizontal band as the plot is rotated. If any other pattern is apparent, the candidate summary plot does not contain all the information about the distribution of $y|x$, implying that h is noticeably different from $c\beta^T x$. For full confidence that the candidate summary plot $\{h, y\}$ is adequate, a horizontal band of points should be observed for a series of slices that covers the range of h.

This basic checking procedure is too time consuming to be of much practical value, so we suggest a simpler procedure based on a single 2D view. The horizontal axis of this single view is the linear combination of the predictors that is uncorrelated with h. Call this linear combination h_u, and call the plot $\{h_u, y\}$ an *uncorrelated 2D view*. Figure 6.4b is the uncorrelated 2D view for the 3D plot in Figure 6.4a. The points corresponding to the slice in Figure 6.4a are shown in Figure 6.4b. No within-slice patterns are apparent in this slice, or in any other slice, suggesting that $\{h, y\}$ is an adequate summary plot for these data.

Here is how to use the *R-code* to obtain the uncorrelated 2D view for checking a summary plot:

1. Rotate the 3D plot $\{x_1, y, x_2\}$ about the vertical axis to the 2D view $\{h, y\}$ with the strongest linear trend. This view, which is the candidate summary plot, can be remembered for later use by selecting the item "Remember view" from the "Recall/Extract" menu. This plot should be similar to Figure 6.4a.

2. Select the item "Extract Horizontal" from the "Recall/Extract" menu of the candidate summary plot $\{h, y\}$ and give a name to the horizontal screen variable. The default name h will be used for the rest of this discussion, but meaningful names are always better.

3. Select "Extract uncorrelated 2D plot" from the "Recall/Extract" menu. This produces the uncorrelated 2D view $\{h_u, y\}$ in a separate window but does not change the current 2D view $\{h, y\}$ of the 3D plot $\{x_1, y, x_2\}$.

4. Select the item "Slicer. . . " from the plot menu of the uncorrelated 2D view. This item produces a dialog that allows you to create a slide bar that can be used to view slices in a plot. Type the name for the horizontal screen variable that you chose in Step 2 and then chose a "Fraction" and a "Slice Type." In most cases, a fraction of 0.1 or 0.2 and the "Show Only Slice" option will be good choices. When done push "OK."

5. You now have two plots, $\{h, y\}$ and $\{h_u, y\}$, and a slider on the computer screen. Move them so you can see them all at once; if your

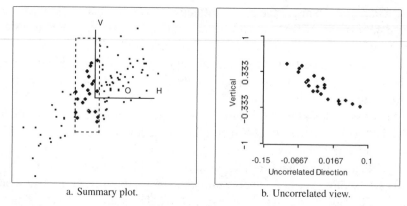

a. Summary plot.   b. Uncorrelated view.

Figure 6.5. A poor candidate summary plot for the "Uncorrelated Predictors, Linear Model" demonstration. Selected points in (a) correspond to the visible slice in (b).

screen is small, you might want to remove the plot controls from the plots using the "Plot Controls" item in the plot's menu. The slider allows you to slice on h using overlapping slices that consist of the fraction of the data you specified in the dialog of Step 4. As you move the slider, the corresponding points will be selected in both plots.

If the points in the uncorrelated 2D view $\{h_u, y\}$ appear as a horizontal band for each slice, then there is no evidence that the candidate summary plot is inadequate. A horizontal band is an indication that y and h_u are independent within the slice. The points in $\{h_u, y\}$ may move up or down as the slider is moved, but if each slider position produces a horizontal band, then there is no evidence that the candidate summary plot is inadequate. Otherwise, try rotating the 3D plot to obtain another candidate summary plot, and return to Step 1.

Draw the plots equivalent to Figure 6.4, and verify that within each slice in the summary plot the corresponding points in the uncorrelated view form a horizontal band of points.

What happens if h misses important information? Rotate Figure 6.3 away from the strongest linear trend, and use the resulting 2D view as your candidate for the summary plot. One possibility is shown in Figure 6.5a. The corresponding uncorrelated 2D view of this plot is shown in Figure 6.5b. The linear trend in the highlighted points of Figure 6.5b is evidence that the h in Figure 6.5a misses information because the mean of y changes as a function of another combination of the predictors.

To summarize, assume that the linear regression model (6.1) is appropriate. We can obtain a visual fit of the model by rotating the 3D plot

$\{x_1, y, x_2\}$ to the best static 2D view, the view with the strongest linear trend. This 2D summary plot should capture all the information on y that is available from x. Inspecting slices of the corresponding uncorrelated 2D view can determine if there may be a better 2D summary plot. If no linear or nonlinear trends appear within slices of the uncorrelated 2D view, then the 3D plot can be abandoned in favor of the summary plot.

The variable h that determines the horizontal axis of the summary plot is a linear combination of the predictors, say $h = b_1 x_1 + b_2 x_2$, where b_1 and b_2 depend on scaling and on the rotation angle. The values of b_1 and b_2 will generally differ from the ols estimates $\hat{\beta}_1$ and $\hat{\beta}_2$, but if h is visually well determined, the ratio b_1/b_2 will be close to $\hat{\beta}_1/\hat{\beta}_2$. In addition, the fitted values from the simple linear regression of y on h will be very close to the fitted values from the linear regression of y on x_1 and x_2.

6.3   CORRELATED PREDICTORS

Finding the strongest linear trend can be quite hard if the predictors are highly correlated. As we have seen in Section 5.5, plots with correlated predictors are difficult to study because of the loss of resolution in potentially important 2D views. A solution once again is to use the "O to e(O|H)" button to make the predictors on the horizontal axes uncorrelated. An example of this is given in Exercise 6.6.

6.4   DISTRIBUTION OF THE PREDICTORS

6.4.1   Nonlinear Predictors

In the examples so far, every 2D view of a 3D plot has shown a linear relationship, and only the strength of the relationship changed as we altered the view. Some 2D views of a 3D plot can show clear nonlinear relationships, however, even when $E(y|x)$ is a linear function of x. The relationship between the predictors x_1 and x_2 is the key. If the regression functions $E(x_1|x_2)$ and $E(x_2|x_1)$ are both linear, then all 2D static views of $\{x_1, y, x_2\}$ will be linear. If either is nonlinear, some static 2D views of the 3D plot $\{x_1, y, x_2\}$ may exhibit a nonlinear trend, even when the linear regression model (6.1) holds.

Select the item "Nonlinear Predictors, Linear Model" from the "Demos:3D" menu, and set up the model with predictors x_1 and x_2 and response y. This again uses artificial data, with x_1 standard normal and $x_2 = x_1^2 + N(0, 0.25)$. As in the previous examples, y is computed as

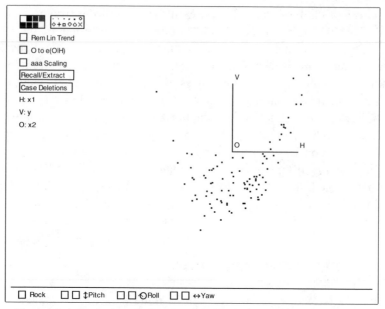

Figure 6.6. A 3D plot from the "Nonlinear Predictors, Linear Model" demonstration.

$y = 1 + 2x_1 + 3x_2 + N(0, 1)$. The linear model holds so all the data points scatter about a plane in three dimensions. The only change between this and the previous versions is that the relationship between the predictors is now nonlinear. The initial 2D view of the plot $\{x_1, y, x_2\}$ is given in Figure 6.6. This initial view is the partial response plot $\{x_1, y\}$ which can be used to estimate the *partial regression function* $E(y|x_1)$. It is clearly nonlinear. Nevertheless, while rotating the 3D plot by using one of the "Yaw" buttons, the points are seen to lie near a plane, as required by the linear model. After returning the plot to the "Home" position, use a "Pitch" button to display the plot of $\{x_1, x_2\}$, which is curved.

We have encountered a curious situation: $E(y|x) = E(y|x_1, x_2)$ is a *linear* function of x while $E(y|x_1)$ is a *nonlinear* function of x_1. To understand this situation, we need the connection between the full regression function $E(y|x)$ and the partial regression function $E(y|x_1)$. We can obtain $E(y|x_1)$ from $E(y|x)$ by taking the expectation of the latter with respect to x_2 while holding x_1 fixed at a value, say $\tilde{x}_1$. More carefully, we obtain $E(y|x_1 = \tilde{x}_1)$ by averaging $E(y|x_1 = \tilde{x}_1, x_2)$ over the values of x_2 that are possible while holding $x_1 = \tilde{x}_1$. This means that $E(y|x_1)$ is the average value of

$$E(y|x) = \beta_0 + \beta_1 x_1 + \beta_2 x_2$$

obtained while holding x_1 fixed:

$$\begin{aligned}
E(y|x_1) &= E[E(y|x)|x_1] \\
&= E[(\beta_0 + \beta_1 x_1 + \beta_2 x_2)|x_1] \\
&= \beta_0 + \beta_1 x_1 + \beta_2 E(x_2|x_1) \quad\quad (6.4)
\end{aligned}$$

This equation shows that $E(y|x_1)$ depends on x_1 *and* on $E(x_2|x_1)$, the regression function for x_2 on x_1. This should clarify the example: we constructed the example to have $E(x_2|x_1) = x_1^2$, and so

$$\begin{aligned}
E(y|x_1) &= \beta_0 + \beta_1 x_1 + \beta_2 E(x_2|x_1) \\
&= 1 + 2x_1 + 3x_1^2
\end{aligned}$$

The regression curve for y as a function of x_1 is therefore quadratic, as can be observed in Figure 6.6.

Rotate the 3D plot to display the partial response plot $\{x_2, y\}$. This plot is approximately linear. Is this expected? Interchanging the roles of x_1 and x_2 in (6.4), we find

$$E(y|x_2) = \beta_0 + \beta_1 E(x_1|x_2) + \beta_2 x_2$$

But $E(x_1|x_2)$ is zero in this example since for each value of x_2, x_1 has a distribution that is symmetric about zero. Thus, $E(y|x_2) = \beta_0 + \beta_2 x_2$, and a linear plot should be expected.

6.4.2 Linear Relationships between Predictors

Now suppose that the regression function $E(x_2|x_1)$ is linear in x_1, so for some constants α_0 and α_1,

$$E(x_2|x_1) = \alpha_0 + \alpha_1 x_1$$

Substituting this into (6.4) gives

$$\begin{aligned}
E(y|x_1) &= \beta_0 + \beta_1 x_1 + \beta_2 E(x_2|x_1) \\
&= \beta_0 + \beta_1 x_1 + \beta_2(\alpha_0 + \alpha_1 x_1) \\
&= (\beta_0 + \beta_2\alpha_0) + (\beta_1 + \beta_2\alpha_1)x_1 \quad\quad (6.5)
\end{aligned}$$

The partial regression function $E(y|x_1)$ will then be linear, although the slope and intercept in the plot of $\{x_1, y\}$ cannot be used to estimate β_0 or β_1.

As an example, suppose that $\beta_1 = 0$, which means that x_1 is not needed in the regression of y on x. What can we expect to see in the partial response plot $\{x_1, y\}$? From (6.5) we see that

$$E(y|x_1) = (\beta_0 + \beta_2\alpha_0) + \beta_2\alpha_1 x_1$$

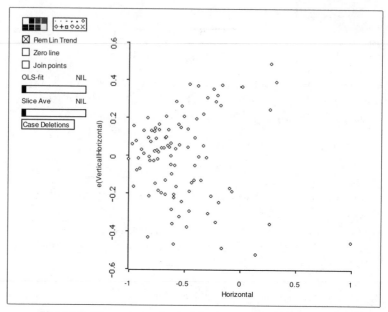

Figure 6.7. Nonconstant variance arising because of nonlinear predictors.

Although x_1 is not needed in the full regression, the partial response plot $\{x_1, y\}$ may show a linear trend if $\beta_2\alpha_1 \neq 0$.

6.4.3   Partial Variance Functions

So far we have seen how nonlinear relationships between the predictors can influence the behavior of partial regression functions. Let's next consider *partial variance functions*. Again rotate the 3D plot in Figure 6.6 to the partial response plot for x_2, $\{x_2, y\}$. Extract this 2D view by using the item "Extract 2D Plot" from the "Recall/Extract" pop-up menu. The extracted 2D plot shows a linear trend as we saw before, but it seems to exhibit nonconstant variance as well, with the points at the right of the plot more spread out than the points at the left of the plot. This can be seen a bit better by pushing the "Rem Lin Trend" button to remove the linear trend. The result is shown in Figure 6.7. The nonlinear relationship between the predictors causes the apparent dependence of var$(y|x_2)$ on x_2.

To summarize, the behavior of the partial response plot $\{x_1, y\}$ depends on the regression function $E(x_2|x_1)$, even when the linear model (6.1) holds. The regression function for this plot will be linear if $E(x_2|x_1)$ is linear in x_1, and the linear model holds. Similarly, the regression function for the partial response plot $\{x_2, y\}$ depends on $E(x_1|x_2)$, which is the regression function for the scatterplot $\{x_2, x_1\}$. It is possible for $E(x_2|x_1)$ to be linear while $E(x_1|x_2)$ is nonlinear if the relationship between the predictors is

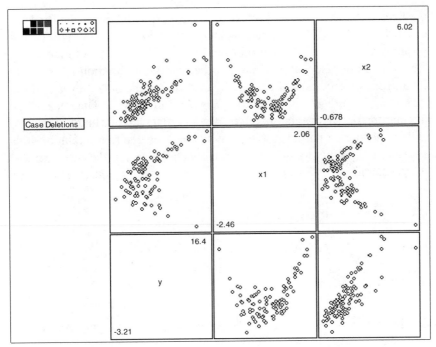

Figure 6.8. Scatterplot matrix for the "Nonlinear Predictors, Linear Model" demonstration.

not monotone, as illustrated by the example of this section. A relationship is *monotone increasing* if an increase in one variable always corresponds to an increase in the other and *monotone decreasing* if an increase in one variable implies a decrease in the other.

6.4.4   Scatterplot Matrices

Figure 6.8 gives a scatterplot matrix of the response and predictors for the "Nonlinear Predictors, Linear Model" demonstration. The bottom row of the figure gives the partial response plots. Both partial response plots exhibit nonconstant variance, and $\{x_1, y\}$ exhibits nonlinearity. We know, however, that $E(y|x)$ is linear in x, but the scatterplot matrix nevertheless makes the problem seem much more complicated. As pointed out at the end of Section 4.4, this illustrates that nonlinear relationships between the predictors makes direct interpretation of partial response plots difficult or impossible.

6.4.5   Multiple Regression

The discussion has been confined to regression problems with two predictors, but the conclusions apply equally to regression problems with many predictors. To illustrate, let's return to the scatterplot matrix of the Big Mac data in Figure 4.2. The bottom row of the scatterplot matrix gives the partial response plots for each of the four predictors. Two or three of them show clear curvature. Does this mean that the regression function for *BigMac* on all four predictors will be nonlinear in the predictors? Several of the 12 plots in Figure 4.2, not including *BigMac*, are curved. This means that the untransformed predictors in the Big Mac data are not linearly related, and so the curved partial response plots could be due to the relationships between the predictors, not to the relationship between the response and the predictors.

6.5 LINEAR PREDICTORS

Using *linear predictors* is a central theme in this book. A vector $x = (x_1, \ldots, x_p)^T$ of p predictors is a set of linear predictors if all regression functions of the form

$$E(a_0^T x | a_1^T x, \ldots, a_k^T x)$$

are linear functions of $a_1^T x, \ldots, a_k^T x$, where $a_0, \ldots, a_k$ are any $p \times 1$ vectors of constants. In other words, the regression function for regressing any linear combination of the predictors $a_0^T x$ on any other set of linear combinations $a_1^T x, \ldots, a_k^T x$ must be linear. For any graph $\{a_1^T x, a_0^T x\}$, the regression function $E(a_0^T x | a_1^T x)$ must be linear. If the predictors are normally distributed, then these conditions are satisfied, although normality is a stronger condition than is actually needed.

Techniques requiring linear predictors can still be applied under modest violations of the assumption and may still yield practically useful results. The checks for linear predictors using scatterplot matrices and 3D plots of predictors described in this book are adequate for most practical applications, although they do not guarantee that the conditions are satisfied.

We have seen in Section 4.3 that all the bivariate relationships between the predictors in the Big Mac data could be approximately linearized by replacing the predictors by their logarithms. A further check on the assumption of linear predictors is to examine 3D plots with predictors on all three axes. After removing the linear trend and changing from "O to e(O|H)" these plots should appear as a circular point cloud with no clear nonlinear trends. If they do, then the assumption of linear predictors is probably reasonable. Try examining 3D plots of predictors in log scale for the Big Mac data. Given the small sample size, and possibly excluding the point for Lagos, nothing in these plots excludes the possibility of linear predictors.

We can now examine the partial response plots in the last row of Figure 4.4. These partial response plots are all curved. Assuming the log-transformed predictors in the Big Mac data are linear predictors, we will eventually be able to conclude that the regression function for *BigMac* must be curved. We will return to this aspect of the Big Mac data in Chapter 10, and we will see that the linear model does adequately describe these data if we allow transformation of the response as well as of the predictors.

6.6   COMPLEMENTS

The material in this chapter is drawn partly from Cook (1994). Procedures for checking summary plots, including the method in Section 6.2.4, are investigated by Cook and Wetzel (1993). Eaton (1986) has shown that the conditions given in Section 6.5 are satisfied if and only if x follows an elliptically contoured distribution. The data for Exercises 6.7 and 6.8 are taken from Weisberg (1985).

EXERCISES

6.1. Suppose model (6.1) holds. Under what conditions is the partial regression $E(y|x_2)$ linear? What plot can help to decide if the required condition is satisfied?

6.2. Select the item "Correlated, X2 not needed, Linear Model" from the "Demos:3D" menu and specify a regression with response y and predictors x_1 and x_2. These data were generated with two highly correlated normal predictors, $corr(x_1, x_2) = 0.95$ and response $y = 1 + 2x_1 + N(0, 1)$ so x_2 is not needed.

6.2.1.   Construct a scatterplot matrix of the response and the two predictors. That the partial response plot $\{x_1, y\}$ shows a linear trend is unsurprising, since x_1 is the important predictor. Explain why the partial response plot $\{x_2, y\}$ shows a strong linear trend as well. Describe how you might construct an example so that the same model holds but the partial response plot $\{x_2, y\}$ shows a strong nonlinear trend.

6.2.2. Construct the 3D plot $\{x_1, y, x_2\}$ and then push the "O to e(O|H)" button. The resulting plot shows that only x_1 is relevant, which is the correct conclusion in this case. Why does this work? Would it work if you plotted $\{x_2, y, x_1\}$?

6.3. Return to the "Uncorrelated Predictors, Linear Model" regression from the "Demo:3D" menu and construct the plot $\{x_1, y, x_2\}$. Use the left

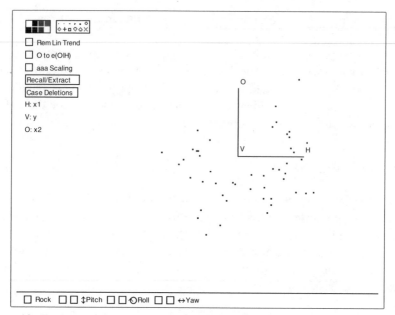

Figure 6.9. The "Uncorrelated predictors, linear model" demonstration with a block of points removed.

"Pitch" button to rotate to the plot $\{x_1, x_2\}$. Select points in the top middle of the point cloud, and then remove the points using the "Remove Selection" item of the plot's menu until the pattern remaining is a clear U-shape. Your plot should now look like Figure 6.9. Removing the points made the relationship between the remaining values of x_1 and x_2 clearly nonlinear but did not change the relationship between y and the two predictors. Suppose now that the values of the predictors in this figure are the only ones that we observed, along with the corresponding responses. We selected a subset of the values of the predictors for analysis, but this does not alter the fact that linear model (6.2) with $(\beta_1, \beta_2) = (2, 3)$ is still the model used to generate the reduced data. Only the particular values of x at which $y|x$ has been observed have changed.

Select "Recall Home" from the "Recall/Extract" menu, and explain why the initial 2D plot is curved. Is there evidence that the regression function is curved?

6.4. Select the item "Nonlinear Predictors, Linear Model" from the "Demos:3D" menu. Construct the 3D plot $\{x_1, y, x_2\}$ and then push the "Rem Lin Trend" button. Explain the contents of this plot and what it means.

6.5. This problem uses data from a study of the growth of children born in Berkeley, California, in 1928–1929 as in Exercise 5.5. We will use data on girls and the three variables *HT9*, *HT18*, and *WT9*, which are, respectively, the child's height at ages 9 and 18 and weight at age 9.

6.5.1. Load the file BGSgirls.lsp from the R-data folder. Use the regression dialog to create a linear model with *HT18* as the response and *HT9* and *WT9* as predictors. Leave the name of the regression as "BGS-girls." Draw the 3D plot {*HT9*, *HT18*, *WT9*} by selecting the "Plot of..." item from the "BGS-girls" menu. After observing the rotating plot, write a short description of the plot.

6.5.2. Obtain a visual fit to this plot by rotating to the 2D view with the strongest linear trend. Use the "Print Screen Coordinates" item on the "Recall/Extract" menu to find the linear combination of the age 9 variables that correspond to the horizontal axis. Compare the ratio b_1/b_2 from the screen coordinates to the estimates $\hat{\beta}_1/\hat{\beta}_2$ from the printed regression output.

6.5.3. Rotate the plot to the best view, and again select the "Print Screen Coordinates" item. Without rotating the plot, select the "Extract Horizontal" item from the "Recall/Extract" menu to create a variable called *h1*. Now compute the regression with *h1* as the response and *HT9* and *WT9* as predictors using the "New model..." item in the regression menu. Verify that the fitted regression is the same as given by the "Print Screen Coordinates" item.

6.5.4. Next, create a regression model with *HT18* as the response and *h1* as the predictor. (Again, use the "New model..." item from the regression menu.) In the standard regression dialog, name this regression *h-model*. How do you think the fitted values from the regression of *HT18* on *HT9* and *WT9* will compare to the fitted values from the regression of *HT18* on *h1*? After thinking about this, draw the plot of one set of fitted values versus the other. You can do this by typing

```
(def plot (plot-points (send bgs-girls :fit-values)
                       (send h-model :fit-values)))
```

In this statement, bgs-girls is the name of the original model, and h-model is the name of the model just created. This plot will have the original fitted values on the horizontal axis and the h-model fitted values on the vertical axis. You can add plot controls to the plot by typing the message

```
(send plot :plot-controls)
```

6.5.5. Obtain the 2D view of {*HT9, HT18, WT9*} with the strongest linear trend and follow the steps in Section 6.2.4 to decide if your 2D view misses relevant information about the relationship between the response and the predictors.

6.6. Select the "Correlated Predictors, Linear Model" item from the "Demo: 3D" menu. The two predictors in this example are denoted by x_1 and x_2; the response is denoted by y. This example is similar to the demonstration in Section 6.1.2, except that the correlation between the predictors is 0.99 rather than zero. Use the regression dialog to set up the model.

Verify the correlation between x_1 and x_2 by drawing the plot of $\{x_1, x_2\}$ and observing that these points fall nearly on a line. Before drawing the 3D plot $\{x_1, y, x_2\}$, how do you think it will look? Now draw and describe the plot.

Try fitting by eye to this plot. Most static 2D views of the 3D plot show a linear relationship of about equal strength; all that changes is the slope. Why is this? Fitting by eye is highly variable. To improve resolution, use the "O to e(O|H)" plot control. Explain what this control does to the plot, and try to fit by eye again. What do you conclude?

6.7. The data for this problem are from an economic study of the variation in rent paid for agricultural land planted with alfalfa in 1977. Alfalfa is a high protein crop used to feed dairy cows. The unit of analysis is a county in Minnesota, and the data include Y, average rent per acre planted to alfalfa; X_1, average rent paid for all tillable land; X_2, density of dairy cows, number per square mile; X_3, proportion of farmland in the county used as pasture; and X_4, an indicator with value 1 if liming is required to grow alfalfa and 0 otherwise.

Load the file `landrent.lsp` from the `R-data` folder, and choose Y as the response variable and X_1 and X_2 as the predictors. We will not use X_4 in this problem.

6.7.1. Examine the scatterplot matrix of Y, X_1, X_2, and X_3. Does the assumption of linear predictors seem plausible for these data? Why or why not? If the assumption of linear predictors does not seem reasonable, use the transformation controls on the plot to transform the predictors to a set of more nearly linear predictors.

6.7.2. Use the "New Model..." item to transform the predictors as suggested in the scatterplot matrix, and set up a new model with Y as the response and the transformed predictors. Examine the 3D plot of your transformed predictors. What do you conclude?

6.7.3. Draw the ols summary plot $\{\hat{y}, y\}$. What would the summary plot look like if the linear model were true? What does it look like here? What do you conclude?

6.8. Imagine a problem intended to model salary of faculty members given their sex (0 = male and 1 = female) and the number of years in current rank. Suppose the following two true regression functions hold:

$$E(Salary|Sex, Year) = 18065 + 201Sex + 759Year$$
$$E(Salary|Sex) = 24697 - 3340Sex$$

6.8.1. Explain how both of these regression functions could be true at the same time.

6.8.2. Data on salaries from a small Midwestern college are given in `salary.lsp` in the R-data folder. These data produce nearly the same estimated regression functions as the true ones given above. Use the data to explain the differences between the estimated regression functions.

CHAPTER 7

Visualizing Regression without Linearity

The methods in the last chapter show how to use graphs to view and summarize data coming from a linear model. In this chapter, we study how to view and summarize data without assuming a specific model. The basic ideas are nearly the same as those of the last chapter, but the implementation will be different because we will no longer have a linear model to guide our actions.

7.1 GENERAL THREE-DIMENSIONAL RESPONSE PLOTS

Imagine a 3D rotating plot $\{x_1, y, x_2\}$ for a regression problem with two predictors. Without relying on a model, the central theme of this chapter is to understand how the distribution of $y|x$ changes with the value of x. The dependence can be very simple, or it can be complex. The simplest sort of dependence is *independence*.

7.1.1 Zero-Dimensional Structure

If the distribution of $y|x$ does not depend on the value of x, then both the regression function $E(y|x)$ and the variance function $var(y|x)$ remain constant as x changes. How will this be reflected in the 3D response plot $\{x_1, y, x_2\}$? As we rotate about the vertical axis, no systematic patterns should appear in any 2D view and the points in any slice of any 2D view should be distributed similarly. We will then say that the 3D plot exhibits *0-dimensional (OD) structure*. The dimension refers to the number of linear combinations of the predictors required to summarize the dependence of $y|x$ on x. If $y|x$ does not depend on x, we have 0D structure because no linear combination of x provides information about y. With 0D structure, the predictors give no information about the response, and an ideal summary plot is simply a histogram of the response.

7.1.2 One-Dimensional Structure

Suppose there is a systematic pattern in a 3D plot. The pattern could be curved, indicating that the regression function is nonlinear in x. The pattern could be fan-shaped, indicating that the variance function var($y|x$) changes with x. Imagine rotating a 3D plot to the most striking version of the pattern, the clearest curve or fan shape, and let $b^T x$ denote the linear combination of the predictors corresponding to the horizontal screen variable of the resulting 2D view. The value of b could be recovered by using the "Print Screen Coordinates" item on the "Recall/Extract" menu. We now ask if the 2D view $\{b^T x, y\}$ is an adequate summary plot: Does it contain all or nearly all of the information on y that is available from x. We studied this question in Section 6.2.4, assuming a model that is linear in the predictors. We can obtain similar results without any assumptions concerning a model.

This brings us to the next type of dependence. Suppose there is a coefficient vector $\beta^T = (\beta_1, \beta_2)$ for which the distribution of $y|x$ is the same as the distribution of $y|(\beta^T x)$ for all values of x. In other words, the single linear combination $\beta^T x$ gives the same information about the distribution of $y|x$ as would the individual components of x. The 3D plot $\{x_1, y, x_2\}$ and the corresponding regression problem will then be said to have *1-dimensional (1D) structure*, because only one linear combination of the original predictors, namely $\beta^T x$, is needed to extract all of the information from x about the distribution of $y|x$. Since the distribution of $y|(\beta^T x)$ is the same as the distribution of $y|(c\beta^T x)$ for any nonzero constant c, the magnitudes of the elements of β really don't matter. Knowing the linear combination $c\beta^T x$ is statistically equivalent to knowing the linear combination $\beta^T x$. This is just a restatement of an argument used in Section 6.1 without relying on a specific model.

Many of the models in regression analysis exhibit 1D structure. One model is

$$y|x = f(\beta^T x) + \sigma \varepsilon \qquad (7.1)$$

where f is a function that may be known or unknown, σ is the usually unknown error standard deviation that is the same for all errors, and ε is a random variable with mean 0 and variance 1. For this model, the regression function is $E(y|x) = f(\beta^T x)$, and the variance function is constant,

$$\text{var}(y|x) = \sigma^2 \text{var}(\varepsilon) = \sigma^2$$

The graph $\{\beta^T x, y\}$ is an ideal summary plot for this model, because it contains all the information about y that is available from x. The linear regression model (6.2) is a special case of (7.1) that adds the further condition that f is linear, $f(\beta^T x) = \beta_0 + c\beta^T x$.

In most regression models the dependence of the distribution of $y|x$ on x is through the mean, $E(y|x)$, but the dependence on x can be more general. Another 1D structure is

$$y|x = \sigma(\beta^T x)\varepsilon \qquad (7.2)$$

In this equation $\sigma(\beta^T x)$ is a nonnegative function that may have a different value for each value of $\beta^T x$, so each observation can have its own error standard deviation. In (7.2) the regression function is constant, $E(y|x) = 0$ for all x, but the variance function changes with x.

A more general model with 1D structure is

$$y|x = f(\beta^T x) + \sigma(\beta^T x)\varepsilon \qquad (7.3)$$

We will call (7.3) the *1D model*. It includes models (7.1) and (7.2) as special cases, and it is the most general model with 1D structure considered in this book.

For the 1D model, both the regression function and the variance function depend on the single predictor $\beta^T x$. This model is usually described as having a variance function that depends on the mean $E(y|x)$, even though more accurately both the regression function and the variance function depend on the fundamental underlying quantity $\beta^T x$.

With 1D structure, the 2D plot $\{\beta^T x, y\}$ is an ideal summary plot because this one plot contains all the available information about $y|x$. If we knew $\beta^T x$, we would not need a 3D plot at all; the 2D plot would be enough.

How can we decide if 1D structure is appropriate for a particular 3D plot? We encountered this question in Section 6.2, and the solution here is almost the same. Rotate the 3D plot to the 2D view with the most striking pattern. This pattern can be linear or nonlinear or it may indicate changing variance. As before, let $b^T x$ be the horizontal screen variable. If 1D structure is appropriate and $b^T x \approx c\beta^T x$ for some nonzero constant c, then the summary plot $\{b^T x, y\}$ will contain most of the information about y in x. This can be checked by slicing the summary plot and observing the corresponding uncorrelated 2D view, as described in Section 6.2.4. The slices should show a horizontal scatter of points in the uncorrelated 2D view. If not, either $b^T x$ is far from proportional to $\beta^T x$ or the 3D plot is not 1D.

7.1.3   Two-Dimensional Structure

If the uncorrelated 2D view shows dependence, regardless of the choice of b, then we say that the 3D plot exhibits *2D structure*. This means that we need to know the values of both predictors to understand the dependence of $y|x$ on x. The ideal summary plot is now a 3D plot, so all 2D views lose information. This is the most general type of 3D plot.

One model with 2D structure is

$$y|x = f(\beta^T x) + \sigma(\alpha^T x)\varepsilon \qquad (7.4)$$

A linear combination of the predictors $\beta^T x$ is required to specify the regression function, while a different linear combination $\alpha^T x$ is required to specify the variance function; both are needed to understand fully the distribution of $y|x$. Another model with 2D structure is

$$y|x = f(\beta^T x, \alpha^T x) + \sigma\varepsilon \qquad (7.5)$$

The regression function is now 2D: $E(y|x) = f(\beta^T x, \alpha^T x)$ depends on the two linear combinations $\beta^T x$ and $\alpha^T x$.

Load the file demo-3d.lsp from the R-data directory. From the "Demos:3D" menu, select the item "Interaction" and rotate the resulting plot about the vertical axis. Does the plot exhibit 0D, 1D, or 2D structure? There is clearly a relationship between y and the predictors, so the plot has either 1D or 2D structure. Rotate the plot to the most striking 2D view as a candidate for a summary plot. The view we selected is shown in Figure 7.1. Use the procedure outlined in Section 6.2.4 to check for a second dimension. First use the "Extract Horizontal" plot control to save the horizontal screen variable. Then, use the "Extract uncorrelated 2D plot" to get the uncorrelated view. Finally, use the "Slicer..." item in the plot's menu to slice the plot on the extracted horizontal variable. Figure 7.2a shows the summary plot with a slice selected. The points corresponding to this slice are plotted in the uncorrelated view in Figure 7.2b. The points in this slice of the uncorrelated view form a curved pattern, so a second linear combination of the predictors is required. Something similar happens regardless of the candidate for the 2D summary plot, and so we must conclude that the plot has 2D structure. The data for this example

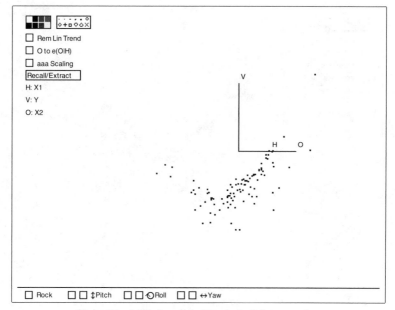

Figure 7.1. A 2D view of the "Interaction" demonstration.

were generated according to the model

$$y = x_1 + x_2 + 2x_1x_2 + N(0, 1) \qquad (7.6)$$

The presence of the interaction term $2x_1x_2$ causes the saddle shape in Figure 7.1 and the 2D structure. There is no single linear combination of x_1 and x_2 that can be used in place of x.

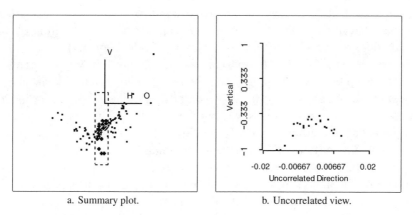

a. Summary plot.   b. Uncorrelated view.

Figure 7.2. The uncorrelated view for the "Interaction" demonstration.

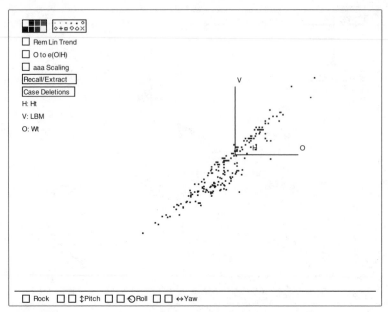

Figure 7.3. Strongest 2D view of the 3D plot {*Ht, LBM, Wt*} for 202 Australian athletes.

7.2   EXAMPLE: AUSTRALIAN ATHLETES DATA

This example concerns modeling *LBM*, lean body mass, as a function of *Ht*, height in centimeters, and *Wt*, weight in kilograms, for a sample of 202 elite Australian athletes who trained at the Australian Institute of Sport. The members of the sample participate in a number of different sports and are about equally split between men and women. Load the file `ais.lsp` from the `R-data` folder. Select *Ht* and *Wt* to be predictors and *LBM* to be the response. The variable *Labels* consists of a combination of the individual's sex and sport. Construct a plot of {*Ht, LBM, Wt*}.

Use the right "Pitch" button to obtain the 2D plot {*Ht, Wt*}. Notice the relationship between the predictors: the points seem to fall in a fan shape, with smallest variation in the lower left and the largest variation in the upper right. What does this say about the relationship between height and weight for these athletes? Select the "Show Labels" item from the plot's menu, and identify the few most extreme points.

Select the "Home" item from the "Recall/Extract" menu, and click the mouse on the plot to make the labels of identified cases disappear. Rotate the plot by holding down the left "Yaw" button. As the plot rotates, choose a 2D view that appears to give the strongest relationship. The view that we selected is shown in Figure 7.3. Because height and weight are highly correlated, determining a single strongest view may be relatively difficult. When you are satisfied, select the "Print Screen Coordinates" item. The ratio of the multipliers for *Ht* and *Wt* will be about 0.35 if your view is close to the ols view.

Using your 2D summary plot, examine the 3D plot for 1D structure. This requires looking at slices in the uncorrelated 2D view as described in Section 6.2.4. Little additional structure is apparent in our plot, so the 2D view shown in Figure 7.3 is a good summary of the full 3D plot. Further analysis of these data need not refer to the 3D plot, as the 2D plot contains nearly all the information.

7.3   EXAMPLE: ETHANOL DATA

This example uses data from an industrial experiment to study exhaust from an experimental one-cylinder engine using ethanol as a fuel. The response variable, which we denote by *NOx*, is the concentration of nitrogen oxide plus the concentration of nitrogen dioxide, normalized by the work of the engine. The units of the response are micrograms of *NOx* per joule. The two predictor variables are *E*, a measure of the richness of the air and fuel mixtures at which the engine was run, and *C*, the compression ratio of the engine. There are a total of 88 observations.

The data can be obtained by loading the file `ethanol.lsp` from the `R-data` directory. In the regression dialog select *C* and *E* to be predictors and set *NOx* to be the response. Construct the plot {*C*, *NOx*, *E*}.

Prior to rotation the static 2D plot on the computer screen is the partial response plot {*C*, *NOx*}. This plot shows that five values of *C* were used in the experiment, and it suggests that the distribution of *NOx*|*C* is only weakly dependent on the value of *C*, if there is any dependence at all. Now rotate the plot by using the left "Yaw" button. A very strong quadratic tendency in the data is immediately apparent, so the structure of the plot is either 1D or 2D. As in the last example, use uncorrelated 2D views to assess candidate summary plots. What do you conclude about the structural dimension of the regression?

Return to the plot {*C*, *NOx*, *E*}. As the plot is rotated, there is one particular view where the right half of the quadratic trend is very clearly determined but the left half is not as well determined and a different view where the left half is clearly determined but the right half is not as well determined. Figure 7.4 shows these two views. Two different linear combinations of the predictors are required to describe the data, so the structure must be 2D.

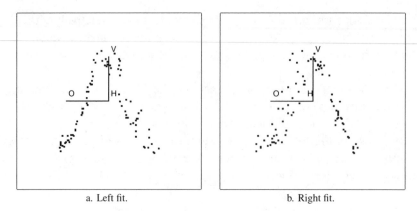

a. Left fit.   b. Right fit.

Figure 7.4. Best half-fits to the ethanol data.

These data can be examined in another way. Begin by rotating to a 2D summary plot that compromises between the views in Figure 7.4. You may need to select the item "Slower" from the plot's menu to get more precise control over the rotation. The plot that we selected is shown in Figure 7.5. The linear combination of C and E on the horizontal axis of the plot in Figure 7.5 can be found by printing the screen coordinates:

```
Linear Combinations on screen axes in current rotating plot.
Horizontal: 2.71893  + -0.0146579 H + 0 V + -2.86916 O
Vertical:   -1.23403  + 0 H + 0.565291 V + 0 O
```

Thus, $h = -0.0147C - 2.869E$ is the single variable that we have chosen visually to explain the variation in the response, NOx. We know from Figure 7.4 that the distribution of $NOx|h$ must still depend on the predictors. Our job is to construct a better visual characterization of the nature of the dependence.

Focus on the data for the largest and smallest values of C in Figure 7.5. Selecting points in a linked plot of $\{C, E\}$ can help find these points in the 3D plot. Use the "Plot of..." item in the regression menu to construct this plot. Remove all observations corresponding to the three middle values of C by selecting these points and then using the "Remove Selection" item in the plot's menu. The corresponding points in the 3D plot will be removed since the two plots are linked. From the remaining observations in the plot $\{C, E\}$ select those corresponding to the largest value of C and give them a symbol, such as a ×. Finally, for visual enhancement select the points corresponding to the highest value of C and with the points selected use the "Extract 2D Plot" item from the "Recall/Extract" menu. The result should look like Figure 7.6.

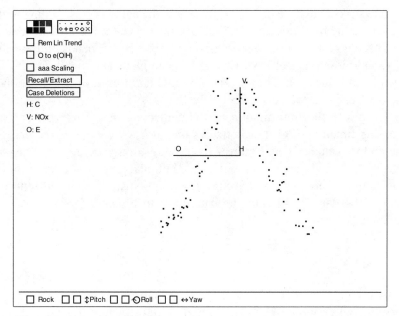

Figure 7.5. Static 2D view of the 3D plot {C, NOx, E} showing the best visual fit overall.

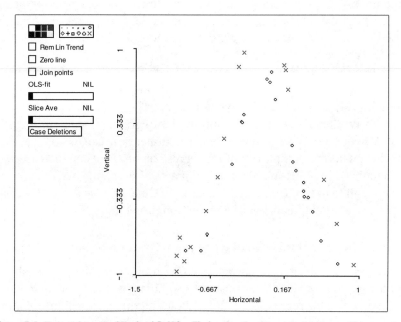

Figure 7.6. Extract from the 3D plot {C, NOx, E} showing the data at the highest and lowest values of C in the best overall 2D view.

One striking feature of Figure 7.6 is that the curve corresponding to the highest value of C lies uniformly above the curve for the lowest value of C. This suggests that the distribution of $NOx|h$ still depends on the value of C and more precisely that $E(NOx|h)$ is an increasing function of C for each fixed value of h. The variable h accounts for most of the variation in the response, but C has an effect as well.

To gain further information on the conjecture that $E(NOx|h)$ is a decreasing function of C at each fixed value of h, we could add the data for the middle value of C to Figure 7.6. If the resulting curve mostly lies between those in Figure 7.6 we will have additional graphical support for the conjecture. If not, we may have information to disprove the conjecture.

Add the data for the middle value of C to Figure 7.6. What do you conclude?

7.4 MANY PREDICTORS

Let $x^T = (x_1, \ldots, x_p)$ be a vector of p predictors. Just as with two predictors, the regression problem is said to have 0D structure if the distribution of $y|x$ does not depend on the value of x. The regression problem is said to have 1D structure if there is a $p \times 1$ vector β so that the distribution of $y|x$ depends on x only through the single linear combination $\beta^T x$. The usual multiple linear regression model with regression function $E(y|x) = \beta_0 + \beta^T x$ has 1D structure. The models given in equations (7.1) and (7.2) have 1D structure when x is $p \times 1$.

If we need two linear combinations of x to describe the distribution of $y|x$, then we have 2D structure, just as we did in the case of two predictors. Models (7.4) and (7.5) are 2D when x is $p \times 1$.

With two predictors the structure of the regression problem must be 0D, 1D, or 2D. With p predictors, the dimension can be $0, 1, \ldots, p$-dimensional. The complexity of the regression problem can increase with the number of predictors, but in many practical problems two dimensions are enough to describe essential features. In Chapter 8 we discuss estimating the structural dimension of a regression problem.

7.4.1 The One-Dimensional Estimation Result

In regression problems with two predictors and 1D structure, we are able to use the 3D plot $\{x_1, y, x_2\}$ to construct a summary plot $\{b^T x, y\}$ that allows us to visualize the regression in 2D without important loss of information. We are also able to use the 3D plot to estimate the structural dimension in cases where dimension is in doubt. This is very useful, but what can be done when we have more than two predictors? The two-predictor methodology will not extend directly to the many-predictor case because the limitations of our 3D world do not allow us to view $(p + 1)$-dimensional plots fully. But good summary plots can be constructed when the predictors are linear.

Consider a regression problem with p predictors $x = (x_1, \ldots, x_p)^T$. If the problem has 1D structure, then for some $p \times 1$ vector $\beta = (\beta_1, \ldots, \beta_p)^T$ the distribution of $y|x$ depends on x only through $\beta^T x$. Suppose we have the 1D model specified in (7.3),

$$y|x = f(\beta^T x) + \sigma(\beta^T x)\varepsilon$$

so both the mean and variance can depend on the linear combination $\beta^T x$. An ideal 2D summary plot would be $\{\beta^T x, y\}$, but this requires knowing β. We don't know β, but it can be estimated up to a proportionality constant. Let

$$\hat{y} = \hat{b}_0 + \hat{b}^T x$$

denote the fitted values from the ols regression of y on x. In performing this regression we are *not* assuming that the linear model holds or even that it yields a sensible fit to the data.

Here is a remarkable result that enables us to construct a good summary plot: Assuming linear predictors, $\hat{b}$ is an estimate of $c\beta$ for some constant c. Since the magnitude of β doesn't matter, the 2D summary plot $\{\hat{b}^T x, y\}$ is an estimate of the ideal summary plot $\{\beta^T x, y\}$. Equivalently, we can take the plot $\{\hat{y}, y\}$ as the summary plot. If the true regression model is the 1D model (7.3), then the summary plot enables us to visualize f and σ. We call this the *1D estimation result*.

With many predictors we require two assumptions — linear predictors and 1D structure — that are not required in the two-predictor case. These assumptions are the price paid for not being able to see in high dimensions.

7.4.2 An Example with a Nonlinear Response

Select the item "Linear Predictors, Nonlinear Model" from the "Demos:3D" menu. In the regression dialog, specify x_1 and x_2 as the predictors and y as the response and then click "Done." A menu titled "Lin/NonLin" will appear on the menu bar. These simulated data comprise 100 observations on two predictors, x_1 and x_2, along with an error ε, all generated as independent $N(0, 1)$ random variables. The response y was computed using

$$y|x = (2 + 2x_1)^2 + \varepsilon$$

so $y|x$ does not depend on x_2, and its dependence on x_1 is nonlinear. This model is of the form given in equation (7.1) with $f(\beta^T x) = (2 + \beta^T x)^2$, $\beta = (2, 0)^T$, and $\beta^T x = 2x_1$. An ideal summary plot is $\{x_1, y\}$. Up to the random errors ε, this plot will recover the quadratic curve, $\{x_1, (2+2x_1)^2\}$. Equivalent ideal summary plots are of the form $\{cx_1, y\}$, for any nonzero value of c, since the graphs will look the same, apart from the labeling of the horizontal axis. One choice of $c \neq 0$ is as good as any other.

The ideal summary plot requires that we know β. Since the model in our example has linear predictors and 1D structure, the summary plot we get from the ols regression of y on x is $\{\hat{b}^T x, y\}$. The plot $\{\hat{b}^T x, y\}$ should look about the same as the plot $\{\beta^T x, f(\beta^T x)\}$. To see if this is in fact the case, select the item "Plot of. . . " from the "Lin/NonLin" menu and construct the 3D plot $\{x_1, y, x_2\}$. Before rotation, the 2D view on the computer screen, $\{x_1, y\}$, is an ideal summary plot. This view shows a J shape that is part of a quadratic curve. The ols plot $\{\hat{b}^T x, y\}$ can be obtained by selecting the item "Recall OLS" from the "Recall/Extract" menu. The plot $\{\hat{b}^T x, y\}$ looks almost the same as the ideal plot $\{x_1, y\}$, as expected.

Let's now consider a second version of this example where the assumption of linear predictors fails. Select the item "Nonlinear Predictors, Nonlinear Model" from the "Demos:3D" menu and set up a regression model as in the first version of this example. The regression menu is called "Nonlin/Nonlin." The data were generated as before, except now $x_2 = x_1^2 + N(0, 1)$. Because of the strong nonlinear relationship between x_1 and x_2, the assumption of linear predictors is no longer appropriate. To see the consequences of this, use the "Plot of. . ." item in the regression menu to construct the 3D plot $\{x_1, y, x_2\}$. The initial 2D view, shown in Figure 7.7a, is again $\{x_1, y\}$. This ideal view shows a J shape, much like the corresponding view in the initial version of this example. Select the item "Recall OLS" from the "Recall/Extract" menu on the plot. The result, shown in Figure 7.7b, is a 2D view with a fairly strong linear trend. We now have two 2D views that give quite different impressions about the data. The ideal 2D view in Figure 7.7a shows a much smaller scatter about its nonlinear trend than does the view in Figure 7.7b. To confirm that the plot of Figure 7.7b misses relevant information, examine slices in the uncorrelated 2D view corresponding to this plot. Within-slice patterns will be visible, confirming that the summary plot suggested by ols misses information.

Here are the essential points of the discussion so far. The ols linear regression of y on p predictors x finds the 2D view, $\{\hat{y}, y\}$ or $\{\hat{b}^T x, y\}$, with the strongest linear trend. How we interpret that view depends on the structure of the data:

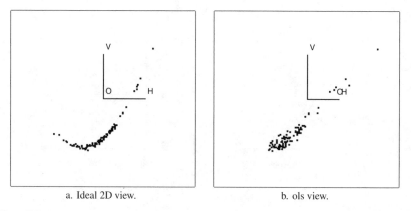

| a. Ideal 2D view. | b. ols view. |

Figure 7.7. Two-dimensional views of a 3D plot from the "Nonlinear predictors/Nonlinear model" demonstration.

- If a linear model $y|x = \beta_0 + \beta^T x + \varepsilon$ is appropriate, then the distribution of the predictors doesn't matter and $\{\hat{y}, y\}$ is a good summary plot. This is the case studied in Chapter 6.

- If the regression function is nonlinear in x but has both 1D structure and linear predictors, then $\{\hat{y}, y\}$ is again a good summary plot.

- If the regression function is nonlinear in x with nonlinear predictors, $\{\hat{y}, y\}$ should not be trusted as a good summary, even if the true model is 1D.

- If the model has more than 1D structure, then the plot $\{\hat{y}, y\}$ must necessarily miss information that may be relevant.

7.5 EXAMPLE: BERKELEY GUIDANCE STUDY FOR GIRLS

The data for this example come from the Berkeley Guidance Study for girls, as described in Exercise 5.5. The data are in the file `BGSgirls.lsp` in the `R-data` folder. The response is *HT18*, height at age 18. The predictors are the height, weight, and strength at age 9, *HT9*, *WT9*, and *ST9*, respectively. The goal is to study how the distribution of the response varies with the three predictors.

Let's first check the assumption of linear predictors. In Section 4.3, this is done using a scatterplot matrix, but with three predictors we can use a 3D plot. Construct the 3D plot {*HT9*, *WT9*, *ST9*} and push the "Rem Lin Trend" and "O to e(O|H)" buttons. This will remove all linear trends from the plot, leaving any nonlinearities behind. If the plot looks like a circular point cloud with no clear nonlinearities, then the assumption

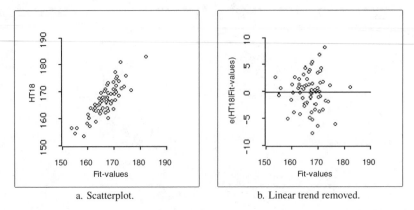

a. Scatterplot.   b. Linear trend removed.

Figure 7.8. Scatterplot of the response versus fitted values from the ols linear regression of *HT18* on *HT9*, *WT9*, and *ST9*, Berkeley Guidance Study for girls.

of linear predictors is reasonable. We didn't see anything in this plot to question the assumption and so we proceed on that basis. Rotate the 3D plot and see if you agree with our conclusion.

Assuming 1D structure, the plot $\{\hat{y}, y\}$ is a good summary plot. Use the "Plot of..." item from the regression menu to construct this plot. The name "Fit-values" will appear in the list of quantities to be plotted, and this item corresponds to $\hat{y}$. The plot shown in Figure 7.8a exhibits strong linearity with no clear evidence of nonlinearity, a conclusion that can be confirmed by using the *lowess* smoother. There is no evidence to indicate that the regression function f of the 1D model (7.3) is not linear.

Strong linear trends in scatterplots can mask secondary features that may be of interest. Remove the linear trend in the plot of $\{\hat{y}, y\}$ to get Figure 7.8b. This plot gives a slight impression that the variance of the response increases with the fitted values. This impression receives some support from the score test for nonconstant variance to be discussed in Section 11.5.

Starting with linear predictors and assuming 1D structure, we have reached the conclusion that a reasonable model for our regression problem is

$$HT18|x = \beta_0 + \beta^T x + \sigma(\beta^T x)\varepsilon$$

where $\sigma(\beta^T x)$ may be an increasing function. Depending on the goals of the study and the type of solution desired, the finding of weak nonconstant variance may not be important.

7.6   EXAMPLE: AUSTRALIAN ATHLETES AGAIN

We modify the previous analysis of the Australian athletes regression described in Section 7.2 by adding the predictor *Sex* to the problem. Specifi-

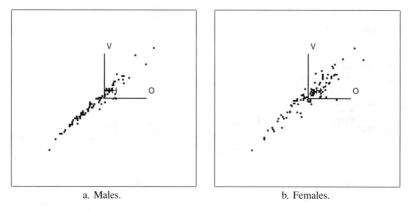

a. Males.   b. Females.

Figure 7.9. Summary plots for the Australian athletes data.

cally, we are interested in characterizing how the distribution of *LBM*|(*Ht*, *Wt*, *Sex*) changes with the values of the three predictors. Since *Sex* takes only two values, *Sex* = 0 for males and *Sex* = 1 for females, this can be done by investigating *LBM* |(*Ht*, *Wt*, *Sex*) separately for males and females. The separate regression functions can be denoted as E(*LBM*|*Ht*, *Wt*, *Sex* = 0) and E(*LBM*|*Ht*, *Wt*, *Sex* = 1) for males and females, respectively.

Set up the regression with response *LBM* and predictors *Ht* and *Wt*. Construct the 3D plot {*Ht*, *LBM*, *Wt*} and a linked histogram of *Sex*. The histogram will serve as a convenient way of focusing on males or females in the 3D plot. In the histogram, select the bar at *Sex* = 0 for males and then from the menu for the 3D plot choose "Focus on Selection" and then "Rescale Plot." The 3D plot now consists of just the data for males and can be investigated as any 3D plot.

Investigate the structural dimension of the 3D plot {*Ht*, *LBM*, *Wt*} for males. What do you conclude? We decided that the plot is adequately characterized by 1D structure; our summary plot is shown in Figure 7.9a. The horizontal screen variable of our summary plot is called h_m. The summary plot suggests that the regression function for males E(*LBM*|*Ht*, *Wt*, *Sex* = 0) is linear in h_m while the variance function for males increases with h_m. The latter conclusion can be seen more clearly by using the "Rem Lin Trend" button on an extracted summary plot. In short, the 1D model (7.3) seems to be suitable for males.

Following the steps for males, investigate next the structural dimension of the 3D plot {*Ht*, *LBM*, *Wt*} for females. Again, what do you conclude? Our solution for females is similar to that for males. The structural dimension is 1 and the variance function increases with h_f. The summary plot for females is shown in Figure 7.9b. Figure 7.10 is the same plot as Figure 7.9b, after using the "Extract 2D Plot" plot control, removing the

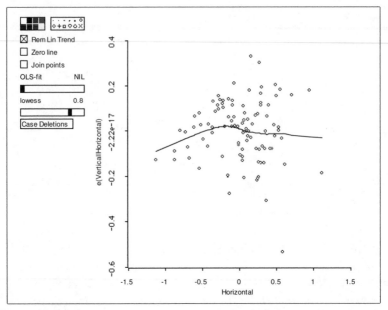

Figure 7.10. Extracted and detrended summary plot for females in the Australian athletes data.

linear trend, and adding a smooth to the graph. Nonconstant variance is evident in this plot and the smooth gives a hint of curvature.

So far we have seen that the data for males and females follow the same general structure. We next investigate if the individual regressions depend on different linear combinations of *Ht* and *Wt*. That is, are h_m and h_f related? If so, we could conclude that the male and female regressions depend on the *same* linear combination of *Ht* and *Wt*. If not, then different linear combinations are necessary and the problem becomes more complicated. The plot $\{h_m, h_f\}$ is shown in Figure 7.11. This plot can be obtained from the "Plot of..." item in the regression menu if you have been following on a computer, and have extracted the horizontal variables in Figure 7.9. Clearly, h_m and h_f have a very strong linear relationship, so we conclude that the regressions for males and females depend on the same linear combination h of *Ht* and *Wt*. Separate models that reflect our findings can be written as

$$LBM|(Ht, Wt, Sex = 0) = \alpha_{0m} + \alpha_{1m}h + \sigma_m(h)\varepsilon \qquad (7.7)$$

and

$$LBM|(Ht, Wt, Sex = 1) = \alpha_{0f} + \alpha_{1f}h + \sigma_f(h)\varepsilon \qquad (7.8)$$

for males and females, respectively. If we assume that $\sigma_f(h) = \sigma_m(h) = \sigma(h)$, then these two equations can be easily combined into one. After a

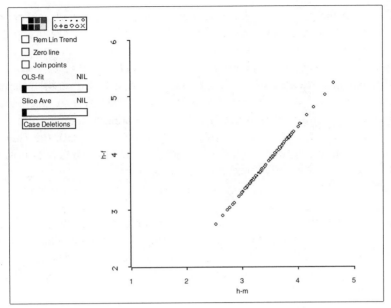

Figure 7.11. Plot of h_f versus h_m in the Australian athletes data.

bit of algebra and changing notation (see Exercise 7.6),

$$\begin{aligned}
LBM|(Ht, Wt, Sex) \;=\;\; & \beta_0 + \beta_1 Ht + \beta_2 Wt + \beta_3 Sex \\
& + \gamma\, Sex \times [\beta_1 Ht + \beta_2 Wt] \\
& + \sigma(\beta_1 Ht + \beta_2 Wt) \times \varepsilon
\end{aligned} \tag{7.9}$$

where the β's and γ are unknown parameters. Model (7.9) is a nonstandard model because it is nonlinear in the parameters.

7.7 COMPLEMENTS

7.7.1 Linearity

The word *linear* is used in many different but related contexts in statistics and in this book. A parameterized regression function can be characterized as linear or nonlinear in the predictors and as linear or nonlinear in the parameters. The regression function

$$E(y|x) = \beta_0 + \beta^T x$$

is linear in both the parameters β_0 and β and in the predictors x. The regression function

$$E(y|x) = \beta_0 + \beta_1 x_1 + \beta_2 x_2 + \beta_3 x_1 x_2$$

is linear in the parameters but nonlinear in the predictors $x^T = (x_1, x_2)$. The regression function corresponding to model (7.9) for the Australian athletes data is nonlinear in the parameters because of the terms $\beta_1\gamma$ and $\beta_2\gamma$, but it is linear in the continuous predictors Ht and Wt.

Statistical software usually distinguishes models based on their parameters. Models that are linear in the parameters can be estimated using standard regression software like the *R-code*. Models that are nonlinear in the parameters require other software and cannot be fit with the *R-code*. Fitting a regression function that is nonlinear in the parameters almost always requires an iterative numerical procedure, while fitting a regression function that is linear in the parameters can be done without iteration. In the general statistical literature the phrases *linear regression* and *nonlinear regression* usually refer to how the parameters enter the regression function.

Fitting the regression function for the Australian athletes data (7.9) requires a nonlinear regression program, but the interpretation is straightforward because the regression function depends on the continuous predictors only through a single linear combination.

7.7.2   ols Summary Plots

An important conclusion of this chapter is that the ols plot $\{\hat{y}, y\}$ is a good summary when we have linear predictors and 1D structure. Without linear predictors there is no guarantee that the ols summary plot will be a good estimate of the ideal summary plot. Good summary plots may sometimes be constructed with nonlinear predictors by using weighted least squares. The general idea is to choose the weights so that the weighted predictors are linear. Methodology for this is developed in Cook and Nachtsheim (1994).

The ols summary plot can also fail if the regression function is symmetric. In this case, the ols estimate $\hat{\beta}$ will estimate zero. For further discussion of such occurrences, see Cook and Weisberg (1991a), Cook, Hawkins, and Weisberg (1992), and Exercise 7.10.

7.7.3   References

The material in this chapter is drawn largely from Cook (1994). The result in Section 7.4.1 that ols estimates can give a consistent estimate of $c\beta$ even when the model is nonlinear is due to Li and Duan (1989) using earlier work by Brillinger (1983). As we will see in Chapter 13, a few unusual points can strongly influence ols estimates. When a few such points are observed in the data, the result $\hat{b}^T x \approx c\beta^T x$ need not hold.

The data for the ethanol example are discussed by Brinkman (1981).

EXERCISES

7.1. From the "Demo:3D" menu, select the item "View a surface." Is the resulting plot 0D, 1D, or 2D?

7.2. From the "Demo:3D" menu, select "Colinearity hiding a curve." Push the "O to e(O|H)" button on the plot. What is the dimension of the structure in the plot? Now remove the linear trend and answer the same question again.

7.3. Repeat the example of Section 7.2, but push the "O to e(O|H)" button before beginning to look at individual slices for 1D or 2D structure. Do your findings change? Also, in the "Slicer. . ." dialog, try setting the fraction to 0.2 rather that 0.1; this will put more data in each slice.

7.4. Suppose we have a regression problem with p linear predictors and a structure that is at most 1D. Let $\hat{y}$ denote the fitted values from the ols regression of y on x. If the plot $\{\hat{y}, y\}$ appears as a random scattering of points with no clear systematic features, what would you choose as the structural dimension of the regression? Why?

7.5. Return to Figure 7.1 and the model (7.6) used to generate the data. We have seen that 2D is needed to describe this figure. Find the two linear combinations needed, and justify your choice. Suppose instead we could view a 4D plot with y on the vertical axis and the three predictors x_1, x_2, and $x_1 x_2$ on the "horizontal" axes. Would this plot continue to show 2D structure? Why or why not?

7.6. *7.6.1.* Show the correspondence between the α's in (7.7) and (7.8) and the β's and γ in (7.9). The easy way to do this is to set $Sex = 0$ to get a correspondence to (7.7) and $Sex = 1$ to get the correspondence to (7.8). Why is model (7.9) nonlinear in the parameters?

7.6.2. How is the regression function in model (7.9) different from

$$E(LBM|(Ht, Wt, Sex)) = \delta_0 + \delta_1 Ht + \delta_2 Wt + \delta_3 Sex$$
$$+ \delta_4 Ht \times Sex + \delta_5 Wt \times Sex$$

7.7. Repeat the analysis of Section 7.2 using the sum of skin folds *SSF* as the response variable, giving a graphically determined model.

7.8. Consider the haystack data introduced in Exercise 5.4. First, analyze the structural dimension of the problem. Then conduct your analysis under the assumption that the structural dimension is 1. Include a discussion of your findings on nonlinearity and nonconstant variance.

7.9. Consider a regression problem with a response y and three predictors $x = (x_1, x_2, x_3)^T$. Write regression functions that correspond to the following situations. For each, give the structural dimension of the regression function: (i) a standard linear model; (ii) a model that is linear in the parameters but nonlinear in the predictors; (iii) a model that is linear in the predictors but nonlinear in the parameters; (iv) a model that is nonlinear in both the predictors and the parameters.

7.10. In this problem we will construct an example that is almost identical to the example given in Section 7.4.2, except that the regression function is slightly modified. In the text window, type the following three statements:

```
(def x1 (normal-rand 100))
(def x2 (normal-rand 100))
(def e (normal-rand 100))
```

This will create three lists of standard normal random numbers, each of length 100. For this example, we want the regression function to be $E(y|x) = x_1^2$, and we want the variance function to be constant with $\sigma = 1$. We can compute y and start the *R-code* as follows:

```
(def y (+ (^ x1 2) e))
(rcode :data (list x1 x2 e y)
       :data-names '("X1" "X2" "e" "y") :name "reg")
```

Set up the regression with x_1 and x_2 as predictors and y as the response. Every time you do this problem, you will get slightly different answers because the data are generated at random each time. In the computer demonstrations, the same data values are used each time.

7.10.1. Compare the data generated here to the data used in Section 7.4.2. How to the regression functions differ? Are the variance functions the same or different? Is the assumption of linear predictors satisfied by the data you generated?

7.10.2. Examine the 3D plot $\{x_1, y, x_2\}$. What is the structural dimension of this plot? Rotate to the strongest 2D view in the plot. Does it correspond to the way you generated the data? How do you know? Mark this view by selecting the item "Remember view" from the "Recall/Extract" pop-up menu.

7.10.3. According to the results in Section 7.4.1, the ols 2D summary plot for these data should be similar to the best view you found visually. Select the item "Recall OLS" from the "Recall/Extract" pop-up menu. Has ols been successful?

7.10.4. Modify the problem so it is similar to the example in Section 7.4.2. The new regression function is $E(y_1|x) = (1 + x_1)^2$ Create a new response y_1 by the commands

```
(def y1 (+ (^ (+ 1 x1) 2) e))
(send reg :add-data y1 "y1")
```

From the "reg" menu, select the item "New Model..." and choose x_1 and x_2 as predictors and y_1 as the response. Draw the plot $\{x_1, y_1, x_2\}$, and observe that the best 2D view chosen will be very similar to the view chosen by ols.

Explain why ols gave a useful answer with y_1 as the response but did not give a useful answer with y as the response.

CHAPTER 8

Finding Dimension

We saw in the last chapter that *if* we have linear predictors, and *if* we have 1D structure, then the 2D plot $\{\hat{y}, y\}$ summarizes the regression problem. As in Sections 4.3 and 7.5, we can check for linear predictors using scatterplot matrices and 3D plots, and transformations can be used to improve linearity among the predictors. Up to now the assumption of 1D structure has been checked only for two predictor problems. In this chapter, we provide two methods for checking the assumption of 1D structure with many predictors.

8.1 FINDING DIMENSION GRAPHICALLY

We begin with an example, again using the Australian athletes data. Load the file `ais.lsp` from the `R-data` folder. We now use the three predictors *Ht*, *Wt*, and red blood cell count *RCC*. As before, use *LBM* as the response. In Section 7.6 using only *Ht* and *Wt* as predictors, we found that *LBM* depends on the same linear combination of *Ht* and *Wt* for males and for females.

Our first task with the enlarged data set is to examine the assumption of linear predictors using the scatterplot matrix shown in Figure 8.1. Apart from a few straggling points, particularly those with the two lowest values of *Ht* and the four highest values of *Wt*, the assumption of linear predictors seems plausible. The 3D plot $\{Ht, RCC, Wt\}$ with the linear trend removed and the "O to e(O|H)" option leads to the same conclusion. However,

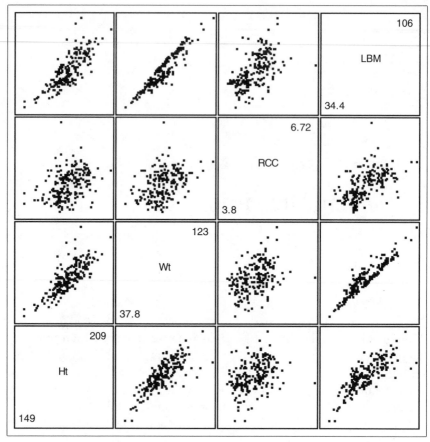

Figure 8.1. Australian athletes data.

any analysis we do should perhaps be repeated with the straggling points removed from the data, as will be illustrated in Section 8.2. A detailed discussion of the role of individual points is given in Chapter 13.

The next question of interest is dimensionality. From the partial response plots, we can eliminate 0D structure, since *LBM* is related to at least one of the predictors. If the structure is 1D, then the plot $\{\hat{y}, y\}$ would be a useful summary plot. The problem may become considerably more complicated if more than 1D structure is needed.

8.1.1 The Inverse Regression Curve

Suppose for a moment that 1D structure is in fact appropriate. This means that the distribution of $y|x$ depends on x only through a single linear combination $\beta^T x$. To see the dependence and verify that it is indeed 1D, we would need to draw a $(p+1)$-dimensional plot of the p predictors and the response, a discouraging prospect if the number of predictors exceeds 2. Suppose we could turn the problem around and, rather than study $y|x$, we could study the *inverse regression* problem of $x|y$. The inverse problem has much simpler structure, since it is a collection of p simple regression problems, $x_1|y, x_2|y, \ldots, x_p|y$, and each of these p problems can be studied with a simple 2D scatterplot.

We are again aided by a remarkable result. Assume linear predictors and 1D structure. Then the regression and variance functions for each of the simple inverse regression problems have the form, for $j = 1, \ldots, p$,

$$E(x_j|y) = E(x_j) + \alpha_j m(y) \tag{8.1}$$

$$\text{var}(x_j|y) \approx \text{var}(x_j) + \alpha_j^2 v(y) \tag{8.2}$$

where α_j is the same in each equation, possibly positive, negative, or zero. These equations require some further explanation. Suppose we examine the p scatterplots of $\{y, x_j\}$, $j = 1, \ldots, p$. These are *inverse partial response plots*, which are the usual partial response plots but with the axes reversed. Equation (8.1) says that for each predictor x_j the inverse regression function $E(x_j|y)$ equals the overall mean $E(x_j)$ plus some unknown function $m(y)$ of y, multiplied by a scale factor α_j. The important point is that the function $m(y)$ does not depend on j. All p plots should have the same shape, apart from linear rescaling. If one of these plots is linear, then all must be linear. If one is J-shaped, then all must be J-shaped, differing only by a constant factor that might change the orientation. If this is not the case, then 1D structure must be abandoned.

Condition (8.2) requires that the inverse variance function $\text{var}(x_j|y)$ be approximately the overall variance $\text{var}(x_j)$ plus α_j^2 times an unknown function $v(y)$ that can depend on y, but not on j. When looking at all p plots, the variability must change in the same way in each of the plots, apart from a linear rescaling. If the variability does not change in the same way, then 1D structure must be abandoned. The only exception to this is predictors for which α_j in (8.1) and (8.2) is zero. To be consistent with 1D structure, plots that show no dependence on y in the inverse regression function ($\alpha_j = 0$) must show no dependence in the inverse variance function, even if the variance is not constant in other plots. We will call (8.1) and (8.2) *checking conditions* for 1D structure since they must be satisfied for each predictor if 1D structure is to hold.

8.1.2 Inverse Partial Response Plots

For the Australian athletes example, $p = 3$ and the three inverse partial response plots are given in the last column of Figure 8.1. To examine these

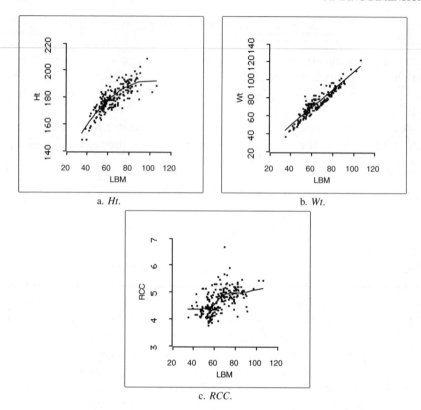

a. *Ht.*   b. *Wt.*

c. *RCC.*

Figure 8.2. Inverse partial response plots for the Australian athletes data.

three plots more carefully, select the item "Inv Partial Response Plots" from the regression menu. This will produce a 2D scatterplot, initially showing the 2D plot with the response *LBM* on the horizontal axis, and one of the predictors on the vertical axis. The only nonstandard feature of this scatterplot is the addition of an extra slide bar. As you push the mouse in this slide bar, the predictor plotted on the vertical axis changes. By pushing repeatedly, you can see all p of the inverse partial response plots.

Fit a smoother to each of the p inverse partial response plots. The smooth in the jth plot is an estimate of the inverse regression function $E(x_j|y)$, which, according to checking condition (8.1), should approximate $E(x_j) + \alpha_j m(y)$ if 1D structure is appropriate. A different smoother can be used for each of the plots, with the goal of obtaining a useful estimate of the inverse regression functions. Any of the methods for simple regression problems can be used to help determine a suitable fit for each of the inverse partial response plots. The three inverse partial response plots are shown in Figure 8.2 with smooths added. The smoother used for *Ht* in Figure 8.2a is a quadratic fit with ols. This plot shows both curvature and variance increasing to the right, but these characteristics are influenced by the straggling points mentioned earlier. The inverse partial response plot for *Wt* in Figure 8.2b appears well matched by a linear fit, and the smooth shown is the ols line. The third inverse partial response plot for *RCC* is in Figure 8.2c. The *lowess* smooth with parameter 0.6 on this plot suggests a nonlinear relationship, with different linear phases for low and high values of *LBM*. The three inverse regression functions are evidently different, so 1D structure is doubtful. Further evidence for this conclusion comes from consideration of variances. The variance of *Ht* increases with *LBM*, but increasing variance is not clearly evident in the other plots. We are again led to the conclusion that 1D structure cannot be supported.

We must now face the difficult question of how to proceed. For this we have no fixed prescription, but we can give a number of guidelines that may help in particular problems. One possibility is to seek additional variables that might remove the necessity for 2D structure. This is the route that we will follow with this example in the next section. Other possibilities include adding cross product or interaction terms to a model, transforming predictors, or separately modeling subsets of the data. These other approaches will be discussed in later examples.

8.2   SLICED INVERSE REGRESSION

One potential problem with the graphical procedure for deciding structural dimension is that we rely heavily on visual impressions. How different must the inverse partial response plots be for us to diagnose at least 2D structure? To answer this question, we turn to a numerical procedure that, assuming normally distributed predictors, can give us a test for 1D structure. The methodology is called *sliced inverse regression*, or *SIR*.

Returning to the Australian athlete example, *SIR* provides a test of the hypothesis that there is a single function $m(LBM)$ and there are constants $\alpha_1, \alpha_2, \alpha_3$ such that

$$
\begin{aligned}
\mathrm{E}(Ht|LBM) &= \mathrm{E}(Ht) + \alpha_1 m(LBM) \\
\mathrm{E}(Wt|LBM) &= \mathrm{E}(Wt) + \alpha_2 m(LBM) \\
\mathrm{E}(RCC|LBM) &= \mathrm{E}(RCC) + \alpha_3 m(LBM)
\end{aligned}
$$

If this were so, then we would expect the smooths from the inverse partial response plots to be the same, except for a scale factor applied to each. To

Table 8.1. Printed Output from *SIR*

```
Sliced Inverse Regression, Number of slices = 28, Response is LBM
Slice sizes are (7 7 7 7 7 7 7 7 7 7 7 7 7 7 7 9 7 7 8 7 8 9 9 7 8 7 7 4)
Std. coef. use predictors scaled to have SD equal to one.
Coefficients                    Lin Comb 1          Lin Comb 2
Predictors                    Raw      Std.       Raw      Std.
Ht                           0.061    0.304      0.016    0.244
Wt                           0.129    0.924     -0.028   -0.633
RCC                          0.990    0.233      0.999    0.734

Eigenvalues                        0.919               0.287
R^2(OLS|SIR lin comb)              0.998               0.998

Approximate Chi-squared test statistics based on partial sums
of eigenvalues times 202

Number of    Test
Components   Statistic       D.F.     p-value
    1          270.6          81       0.000
    2          85.05          52       0.003
    3          27.04          25       0.354
```

get the test statistic, *SIR* uses a smoother based on dividing the data into nonoverlapping slices according to the response y, and then averaging x_j in each slice. The number of slices is the tuning constant for this smoother.

To run *SIR* in the *R-code*, select the item "Inverse Regression" from the regression menu. This will give you a dialog to set a few options. You can choose the number of slices; the default is usually an acceptable choice. You can choose to use either the response or the ols residuals to define the smoother; we will use only the response here. Finally, you can choose the method of analysis. Of the three methods available, only *SIR* is described in this book; references for the other two methods are given in Section 8.5. When you push the "OK" button without changing the default options, after a lengthy calculation you will get a 3D plot and the printed output given in Table 8.1.

The 3D plot produced by this item is of the form $\{h_1, y, h_2\}$, where $h_1 = b^T x$ and $h_2 = a^T x$ are linear combinations of the predictors estimated by *SIR*. If a 1D structure is in fact appropriate, then the 2D view on the computer screen prior to rotation $\{h_1, y\}$ is the summary plot estimated by *SIR*. If a 2D structure is needed, then a second linear combination is necessary. This is the linear combination of the predictors h_2 on the out-of-page axis. Given 2D structure, the full 3D plot $\{h_1, y, h_2\}$ is the summary plot produced by *SIR*. The sample correlation coefficient between h_1 and h_2 is zero by construction.

The coefficient vectors b and a for the linear combinations h_1 and h_2 are given in the first part of Table 8.1 in the columns labelled `Raw`. For the athletes data,

$$h_1 = 0.061Ht + 0.129Wt + 0.990RCC$$

and

$$h_2 = 0.016Ht - 0.028Wt + 0.999RCC$$

Both b and a are normalized to have length 1. The columns marked `Std.` in Table 8.1 give the coefficients that would have been obtained if each of the predictors had been rescaled to have standard deviation 1. In this scaling, all the predictors are equally variable, and the coefficients may be easier to interpret. The standardized coefficients are also normalized to have length 1.

Below the table of coefficients in the printed output are two lines labelled `Eigenvalues` and `R^2(OLS|SIR lin comb)`. The first linear combination always has the largest eigenvalue; the test for the number of dimensions to be described shortly is a function of the eigenvalues. The R^2 measures give a summary of the agreement between the linear combinations h_1 and h_2 chosen by *SIR* and the ols fitted values $\hat{y}$. They are computed as the R^2 in the regression of $\hat{y}$ on the first *SIR* direction h_1 and on the first two *SIR* directions h_1 and h_2. In the example, R^2 with h_1 is 0.998, indicating that the first *SIR* direction and $\hat{y}$ are almost the same.

SIR provides a series of test statistics to help decide on the structural dimension of the regression. Three test statistics are listed in Table 8.1 at the end of the text output. Each statistic should be compared to the percentage points of a chi-squared distribution with the indicated degrees of freedom. The degrees of freedom depends on the number of slices used in getting the smooths for the scatterplots. The resulting p-values are given in the final column of Table 8.1. The first test statistic, 270.6, is used to test the hypothesis that the regression has 0D structure versus the alternative that the structural dimension is at least 1. Since the p-value for this test is 0 to three decimals, we conclude that the dimension is 1 or more. The second statistic, 85.05, is used to test the hypothesis that the regression has at most 1D structure versus the alternative that the structural dimension is at least 2. The p-value for this test is 0.003, so we conclude that the dimension is at least 2. The final statistic 27.04 is used to test the hypothesis that the regression has at most 2D structure versus the alternative that the structural dimension is at least 3. In view of the p-value, it is reasonable to conclude that the regression is at most 2D. All three tests together indicate that the structural dimension of the regression is 2, which is in agreement with the graphical analysis done earlier in this chapter. This means that the 3D *SIR* summary plot $\{h_1, y, h_2\}$ should have 2D structure. Can you find it?

The 3D *SIR* summary plot is dominated by a strong linear trend in $\{h_1, y\}$, nearly obscuring all other information. However, return the plot to the "Home" position, and then use one of the "Yaw" buttons to rotate the plot a few degrees until the 2D view shown in Figure 8.3 or its mirror image is obtained. The points can be seen to fall nearly on two planes. Figure 8.4 is the view of the 3D *SIR* plot $\{h_1, h_2\}$, looking down from the vertical axis. This plot shows that the the planes have little overlap. The second direction is needed to capture this difference between the two planes. The two directions that describe the data appear to be h_1 and a second direction that is a linear combination of h_1 and h_2. *SIR* always chooses uncorrelated directions, but the meaningful directions may be linear combinations of these.

As a next step, we need to understand the two planes. Referring back to the inverse partial response plots in Figure 8.2, a similar jump between two linear trends is apparent in the plot for *RCC*, so perhaps *RCC* alone can be used to explain the differences between the two planes. Brush a linked histogram of *RCC*, and see what happens in Figures 8.3 and 8.4. The marked points in the figures correspond to cases with *RCC* larger than 4.72, the sample mean of *RCC*, and so the difference between the two planes can possibly be modelled as a function of *RCC*. This suggests creating a new variable variable, say $RCCI = 1$ if $RCC > 4.72$, and 0 otherwise, and fitting a model with *LBM* as the response, and with predictors *Ht*, *Wt*, *RCC*, *RCCI*, *Ht* × *RCCI*, *Wt* × *RCCI*, and *RCC* × *RCCI*. The interactions are required to model 2D structure. The variable *RCCI* can be created by using the `cut` function described in Section A.7.2.

In this particular example, we have additional information that might be relevant; in particular, we know that the sample consists of about half females and half males. Because *Sex* and *RCC* are related, it is possible that *RCCI* is serving as a surrogate for *Sex* and that the 2D structure is explained better by sex differences. Fit the interaction model described earlier with *RCCI* and then with *RCCI* replaced by *Sex*. What do you conclude?

If the 2D structure is due to *Sex*, then we could expect 1D structure when analyzing each sex separately. To see if that is the case, let's consider the females only. Draw a histogram of *Sex*, and select all the males, $Sex = 0$, and use the "Case Deletions" plot control to delete all the males

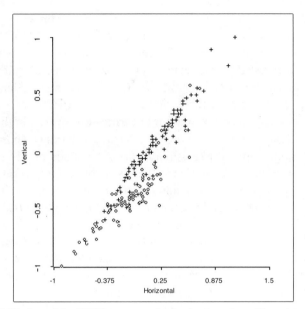

Figure 8.3. A 2D view of the *SIR* 3D summary plot for the Australian athletes data. Highlighting indicates two separate planes in this view.

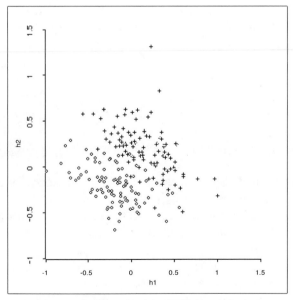

Figure 8.4. Scatterplot of the two directions chosen by *SIR*. Point marking is the same as in Figure 8.3.

from the analysis. Using just the females, examine the three inverse partial response plots and compare them to the plots in Figure 8.2. The biggest difference is in the plot for *RCC*. Since the females have nearly uniformly smaller *LBM* than the males, the points for the females are to the left of the points for males. For females alone, *RCC* is nearly independent of *LBM*, and we estimate the inverse regression function for *RCC* using an ols linear fit. This accounts for one of the nonlinearities discovered in the analysis with the two sexes combined. However, the inverse partial response plot for *Wt* is linear and the plot for *Ht* still appears to be quadratic. To investigate $E(Ht|LBM)$ further, we constructed the ols regression of *Ht* on *LBM* and LBM^2. The coefficient for the quadratic term is large relative to its standard error, supporting the visual assessment of the plot {*LBM, Ht*}. Since the shapes are different, 2D structure may well remain, even for females only.

Apply *SIR* to the data for females only by selecting "Inverse Regression" from the regression menu and using all the defaults. We get a *p*-value for 2D structure of about 0.03, confirming the results from the inverse partial response plots. Your answer may differ slightly because *SIR* depends on the number of slices and the observations put into each slice. This data set has ties among the responses, and the handling of ties affects the slice definitions and hence the answers.

How do we proceed now? From the inverse partial response plots, the 2D structure is due to the curvature in the plot of *Ht*, shown in Figure 8.5. If this plot were straight, then the checking condition (8.1) would be satisfied, and we would have 1D structure. Careful examination of Figure 8.5 draws attention to a few extreme points that contribute substantially to the curvature. Five points are marked with a × on the plot. Much of the need for a curve, rather than a straight line, is contained in these five points. Select these points and delete them using the "Case Deletions" plot control. The smooth fit to the plot becomes nearly straight, the coefficient of LBM^2 is no longer relatively large, and the second *p*-value in the *SIR* output increases to about 0.6, suggesting that 1D structure may be adequate except for a few isolated points.

In summary, the decision between 1D and 2D structure for the *females* in the Australian athletes data is strongly influenced by a few outlying points. Further progress may require additional data to confirm structural dimension. The 2D summary plots {$\hat{y}, y$} or {h_1, y} may well lead to adequate models for many goals, regardless of the structural dimension.

The *SIR* tests require normally distributed predictors. This is more restrictive than the assumption of linear predictors, and the results of the tests can be sensitive to nonnormality. Consequently, the tests should be used only as rough guides to be confirmed by use of the inverse partial response

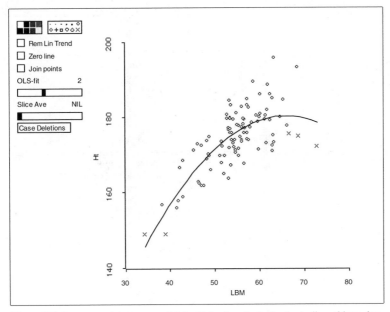

Figure 8.5. Inverse partial response plot for *Ht* for females in the Australian athletes data.

plots. In particular, if the tests indicate a 2D structure that is not evident in the 3D *SIR* plot, it may be best to believe your eyes. No test results will be printed if the number of slices is less than or equal to $p + 1$, although the *SIR* coefficients may still be useful.

8.3 EXAMPLE: ETHANOL DATA REVISITED

Load the file `ethanol.lsp` from the `R-data` folder, which was last analyzed in Section 7.3. In that section, we determined visually that there was a strong, mostly quadratic effect due to E; that C had a small, linear effect; and that 2D structure was necessary to describe the data.

Shown in Figure 8.6 is the scatterplot matrix for the two predictors and the response, *NOx*. Examine first the plots $\{E, C\}$ and $\{C, E\}$ for linearity. Focus on these plots by option-shift-clicking the mouse on each of them. Use the smoother and the "OLS-fit" plot controls to decide if these graphs can be summarized by linear functions or if curves are required. The plot $\{E, C\}$ shows recognizable curvature that may have a little impact on methods requiring linear predictors, but it is unimportant for this example.

Next, turn to the inverse partial response plot $\{NOx, E\}$. This plot is very clearly patterned, but the inverse regression function $E(E|NOx)$ is nearly

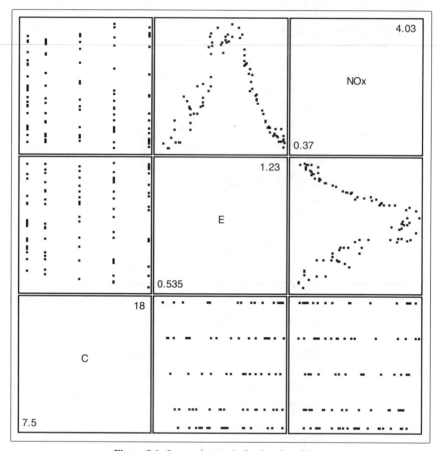

Figure 8.6. Scatterplot matrix for the ethanol data.

a constant function of *NOx*; you can verify this by extracting the plot and adding a smooth. What changes in this plot is the variance: the variability is much larger at the left of the plot than it is at the right. Comparing our summaries of these two plots with the checking conditions (8.1) and (8.2), we can only conclude that this problem has 2D structure, because the variance functions in the two plots are of different shapes. This agrees with our finding from the 3D plot when we analyzed these data before.

Let's turn now to *SIR* by selecting the item "Inverse Regression," again using the defaults. The printed output suggests only 1D structure because the *p*-value for two components is large, equal to 0.345. Next examine the initial view of the 3D plot produced by *SIR*, as shown in Figure 8.7. The plot has completely missed the quadratic dependence of *NOx* on *E* and has only recovered the relatively minor linear dependence of *NOx* on *C*.

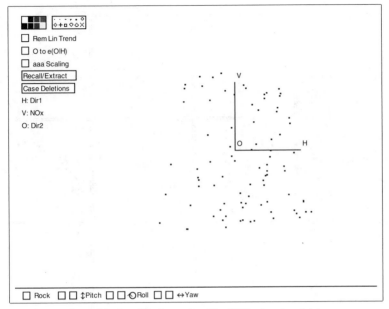

Figure 8.7. Best 2D plot produced by *SIR* for the ethanol data.

The *SIR* methodology only uses the mean checking condition (8.1), not the variance checking condition (8.2). It does not respond to plots with nonconstant variance, and in particular it may miss quadratic dependence like the dependence of *NOx* on *E*, even when the dependence is visually obvious.

8.4 EXAMPLE: AIR QUALITY DATA

The file `air.lsp` in the folder `R-data` gives air quality readings taken on 111 nearly consecutive days in the New York City area in 1973. The response is the Ozone concentration in parts per billion. There are three predictors: *SolR*, solar radiation in Langleys; *Wind*, the wind speed in miles per hour; and *Temp*, the temperature in degrees Fahrenheit. These are not the same data used in Chapter 2 but are from a separate study from a different part of the United States. This fairly complicated example is included to illustrate some of the issues that can arise in an analysis of dimension.

We begin by setting up the regression with *Ozone* as the response and the other variables as the predictors and examining the scatterplot matrix of all four variables, as shown in Figure 8.8. Our first interest is in determining

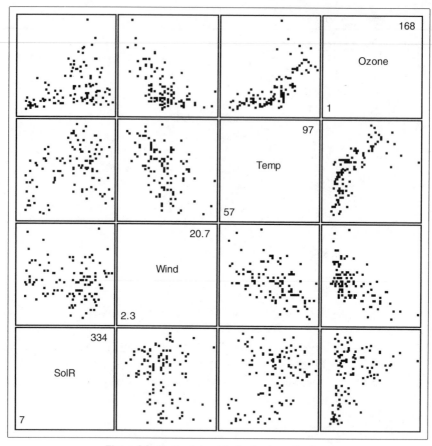

Figure 8.8. Scatterplot matrix for the air quality data.

if the assumption of linear predictors is reasonable or not, and for this we examine all the plots not including the response. The plots {*SolR, Temp*} and {*SolR, Wind*} suggest that the regression functions E(*Temp|SolR*) and E(*Wind|SolR*) are not monotone. If the true regression functions were not monotone, then power transformations of the predictors will not be effective in achieving linearity. Let's examine these two plots more carefully in their own 2D windows, as in Figure 8.9. As expected, both plots suggest some curvature, but the curvature is not very large relative to the variability in the data. Since this problem has only three predictors, we can examine the predictors further by viewing them in a 3D plot, after applying both the "Rem Lin Trend" and "O to e(O|H)" plot controls. No dependence in this plot would imply linear predictors. As we rotate the plot, we see that the 2D view in Figure 8.9a is possibly the worst case of any 2D view, the one with the strongest nonlinearity. Since this nonlinearity is not very strong

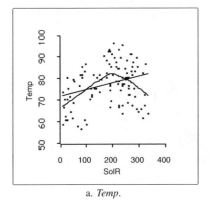
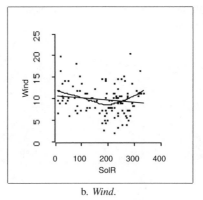

a. *Temp.*   b. *Wind.*

Figure 8.9. Extraction from the scatterplot matrix. In each plot, a *lowess* smooth with parameter $f = 0.6$ and the ols line have been added.

relative to the variation in the data, we proceed with analysis based on the tools of this chapter. Section 8.5 has more on predictor nonlinearities.

We now turn to the question of structural dimension by examining the three plots in the last column of Figure 8.8 with the "Inv Partial Response Plots" item in the regression menu. We added a *lowess* smooth with fraction $f = 0.6$ for each plot. The inverse partial response plots for *SolR* and *Temp* with the smooths superimposed are shown in Figure 8.10; the marked points in these plots will be discussed shortly. The smooths in these plots are both curved and of different shapes. According to the mean checking condition (8.1), we can eliminate 1D structure.

Table 8.2 shows the numerical summary of applying *SIR* with the default number (22) of slices; because of ties in the response, the actual number of slices used is only 19. The p-values given at the end of the *SIR* output in Table 8.2 indicate 1D structure, which is in conflict with the conclusion from the purely graphical analysis. To resolve the situation, we return to the *lowess* smooths in Figure 8.10. The smooths may not be well supported by the data. This suggests comparing the fit of the *lowess* curve to the fit that would be obtained assuming that the regression functions are straight lines. We can do this approximately by fitting a high degree polynomial to approximate the *lowess* fit. We used a fifth degree polynomial, which was reasonably similar to *lowess*. Standard tests of significance suggest that the polynomial fits are better than fitting a straight line to each of the plots. We conclude that the smooths in Figure 8.10 are reasonable.

After further inspecting Figure 8.10, we conjectured that the few points with the largest and smallest values of *Ozone* may substantially account for the curvature in the smooths. The points with the largest seven and smallest

Table 8.2. Printed Output from *SIR* Applied to Air Quality Data with 22 Slices Specified in Dialog

```
Sliced Inverse Regression, Number of slices = 19, Response is Ozone
Slice sizes are (6 7 6 8 5 8 7 5 7 6 7 5 5 5 5 5 5 5 4)
Std. coef. use predictors scaled to have SD equal to one.
Coefficients             Lin Comb 1          Lin Comb 2
Predictors             Raw      Std.        Raw      Std.
SolR                   0.023    0.304      -0.200   -0.902
Wind                  -0.777   -0.400      -0.379   -0.067
Temp                   0.628    0.865       0.904    0.427

Eigenvalues                      0.751              0.167
R^2(OLS|SIR lin comb)            0.984              0.985

Approximate Chi-squared test statistics based on partial sums
of eigenvalues times 111

Number of    Test
Components   Statistic      D.F.    p-value
    1          117.4         54      0.000
    2          33.96         34      0.470
    3          15.4          16      0.495
```

nine values of *Ozone* are marked with special symbols in Figure 8.10. We deleted these 16 points and recomputed the *lowess* smooths using the same fraction. The new smooths show relatively little curvature. Without the 16 extreme values of *Ozone*, the structural dimension appears to be 1, in agreement with the results of *SIR*.

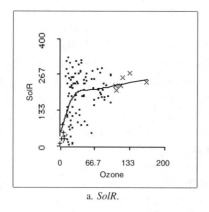

a. *SolR*.

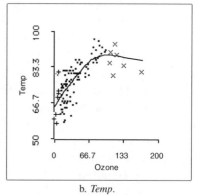

b. *Temp.*

Figure 8.10. Inverse partial response plots for the air quality data with *lowess* smooths and 16 marked points.

Table 8.3. Printed Output from *SIR* Applied to Air Quality Data with 28 Slices Specified in the Dialog

```
Sliced Inverse Regression, Number of slices = 25, Response is Ozone
Slice sizes are (5 4 4 6 4 4 4 5 4 7 5 4 4 5 4 5 4 4 4 4 4 5 5 4 3)
Std. coef. use predictors scaled to have SD equal to one.
Coefficients              Lin Comb 1          Lin Comb 2
Predictors              Raw      Std.        Raw      Std.
SolR                   0.020     0.296     -0.149    -0.927
Wind                  -0.853    -0.497     -0.866    -0.211
Temp                   0.522     0.815      0.477     0.311

Eigenvalues                     0.755               0.312
R^2(OLS|SIR lin comb)           0.996               0.997

Approximate Chi-squared test statistics based on partial sums
of eigenvalues times 111

Number of    Test
Components   Statistic       D.F.    p-value
     1         147.6          72      0.000
     2         63.81          46      0.042
     3         29.2           22      0.139
```

Why did *SIR* miss the extra dimensions due to the extreme values of *Ozone*? The answer seems to be that the slices were too large and the 16 extreme points were not in slices by themselves. We can achieve this by increasing the number of slices. Choosing 28 slices in the *SIR* dialog gives the numerical results shown in Table 8.3. The first two slice sizes are 5 and 4, which account exactly for the nine observations with the smallest values of *Ozone*. The last two slice sizes are 4 and 3, which account for the seven observations with the largest values of *Ozone*. With this slice configuration, *SIR* indicates 2D structure and the *p*-value for 3D structure has decreased noticeably.

We can see what happens in *SIR* if we delete the 16 extreme cases by selecting these points in a graph and then using the item "Delete Selection from Regression" from the "Case Deletions" pop-up menu. If the 3D *SIR* plot is still on the screen, then the printed output will be automatically updated; it should be similar to Table 8.4. Without the 16 cases, the *p*-value for at least one component is very small, while the *p*-value for at least two components is large. Without the 16 extreme cases *SIR* gives a strong indication of 1D structure. Changing the number of slices in *SIR* will not change this conclusion.

Table 8.4. Printed Output from *SIR* Derived from Table 8.3 after Removing Sixteen Cases

```
Sliced Inverse Regression, Number of slices = 22, Response is Ozone
Slice sizes are (4 6 4 4 4 5 4 7 5 4 4 5 4 5 4 4 4 4 5 3 2)
Std. coef. use predictors scaled to have SD equal to one.
Coefficients               Lin Comb 1           Lin Comb 2
Predictors            Raw       Std.        Raw       Std.
SolR               -0.009    -0.120       0.026      0.514
Wind                0.794     0.435       0.980      0.753
Temp               -0.608    -0.892       0.199      0.411

Eigenvalues                   0.732                  0.260
R^2(OLS|SIR lin comb)         0.994                  1.000

Approximate Chi-squared test statistics based on partial sums
of eigenvalues times 95

Number of    Test
Components   Statistic      D.F.    p-value
     1         107.4         63      0.000
     2         37.87         40      0.567
     3         13.15         19      0.831
```

Our conclusions are as follows: The structural dimension is at least 1. The linear combination of the predictors associated with the 1D structure remained reasonably stable throughout our analysis. This can be seen in part by inspecting the coefficients of the first linear combination h_1 in Tables 8.2, 8.3, and 8.4. Inspecting the R^2 values, we see that this linear combination is essentially the same as that given by ols.

The primary evidence for 2D or 3D structure rests with the 16 cases having extreme values of *Ozone*. This is not much information to make a dimension decision, so we cannot expect to resolve the question of dimension with these data alone. From Tables 8.2, 8.3, and 8.4 we see that the second linear combination is not stable. Other predictors not available in the current data might be more effective at explaining the variation at the extremes of the *Ozone* range.

How can we summarize what we have found? We know that the plot $\{\hat{y}, y\}$ will provide a good summary plot with 1D structure, but we have some evidence that the structure is greater than 1D. Regardless, the analysis suggests that the linear combination selected by ols defines one of the dimensions needed. Based on the data at hand, the plot $\{\hat{y}, y\}$ may be a reasonable summary of what we have learned.

8.5   COMPLEMENTS

Duan and Li (1991) initially suggested that the inverse regression $x|y$ can be informative about $y|x$ with 1D structure and proposed *SIR*. Li (1991) showed that inverse regression can be used more generally and provided the test for dimension. Li (1992) provides another method called *pHd* to find inverse structure and a corresponding test for dimension. Cook and Weisberg (1991) suggested a method called *SAVE* that uses the variance checking condition (8.2) to find inverse structure; *SIR* uses the mean checking condition (8.1). All three methods are implemented in the *R-code*. Schott (1994) presents a general method of testing for dimension.

Suppose 1D structure is appropriate. We then know that y depends on x only through a single linear combination $\beta^T x$. If in addition we have linear predictors, we know that the checking conditions (8.1) and (8.2) hold for some $\alpha = (\alpha_1, \ldots, \alpha_p)^T$. In fact α and β are related. If S is the covariance matrix of the predictors, then $\alpha = S\beta$ (Li, 1991). *SIR* uses this result to determine the linear combinations that it obtains.

The assumption of linear predictors is stronger than is really needed for the results of this chapter to hold. The minimal condition is that regression functions $E(b^T x|\beta^T x)$ must be linear in $\beta^T x$ for all b. Although this condition is weaker, it cannot be checked with data because it depends on the unknown β. Minor nonlinearities may be unimportant because of this weaker condition. Hall and Li (1993) have shown that for problems with a large number of predictors the weaker condition will generally be satisfied.

The checking condition (8.2) becomes an equality if the predictors are normally distributed. If they are not normally distributed, this checking condition is not exact, but the approximation is generally quite good.

The air quality data used in Section 8.4 is taken from Chambers, Cleveland, Kleiner, and Tukey (1983). The evaporation data in Exercise 8.4 is taken from Freund (1979).

EXERCISES

8.1. Analyze the males separately in the Australian athletes data, as was done for the females. Qualitatively compare the analyses for the two groups of athletes.

8.2. *8.2.1.* Investigate the regression problem from the Australian athlete data with response *LBM* and predictors *Ht*, *Wt* and a different predictor, *BMI*. Your analysis should include (i) a discussion of the appropriateness of the linear predictor assumption; (ii) application of the general graphical procedure described in Section 8.1.2; (iii) an interpretation of the *SIR* output, including the 3D plot and the test results; and (iv) a summary plot that you think is adequate for the regression, including your rationale. The assumption of linear predictors may or may not be appropriate here. Your discussion should address this issue.

8.2.2.   In the first part of this problem, *SIR* leads to 2D structure that may not be apparent in graphs. Here is a simulation method that can help explain *SIR* in a particular problem.

From previous work, we have decided that the two predictors *Ht* and *Wt* can be treated as linear predictors. First set up a regression with *LBM* as the response and *Ht* and *Wt* as the predictors; name this regression *HtWt*. Next, type the following in the text window:

```
(def y-sim (+ (send htwt :fit-values)
              (* (send htwt :sigma-hat) (normal-rand 202)))))
(send htwt :add-data y-sim "Y-sim")
```

The first of these two commands simulates a new response vector called *Y-sim*, which is equal to the fitted values from the linear regression of *LBM* on *Ht* and *Wt* plus a normally distributed random error with standard deviation equal to the estimated standard deviation from the linear regression; the statement (normal-rand 202) gives a list of 202 standard normal random numbers, and 202 is the sample size in this problem. The next line adds this new variable to the data set. Next, from the "HtWt" menu, select the item "New Model. . . " and set up the model with *Y-sim* as the response and *Ht*, *Wt*, and *BMI* as the predictors. For this model we know the truth: we have 1D structure and *BMI* is irrelevant. Run *SIR* on these simulated data by selecting "Inverse Regression" from the new model's menu. If *SIR* fails to indicate 1D structure, then we must have a problem with the assumptions that must be satisfied for *SIR* to work. What happens here? Can you explain why *SIR* continues to give 2D structure? Repeating this simulation more than once may be helpful.

This simulation method can be used in any problem as a check on *SIR*.

8.3.   Return to the Big Mac data first encountered in Section 4.1. We have previously determined that *Bread*, *BusFare*, *TeachSal*, and *TeachTax* can be made approximately linear predictors by replacing them by their logarithms. You should also use log(*BigMac*) as the response. Although monotonic transformation of the response does not affect the structural dimension of a problem, using logarithms can make the graphs easier to understand. Fit and extract a smooth for each of the inverse partial regression plots, and then examine them in a scatterplot matrix. What do you conclude about dimensionality? Now apply *SIR*. Does *SIR* agree with your graphical solution?

8.4. The file `evaporat.lsp` in the `R-data` folder contains data on daily soil evaporation, *Evap*, for a period of 46 days. There are 10 possible predictors that characterize the air temperature, soil temperature, humidity, and wind speed during a day; use the minimum and maximum of the daily air temperature, soil temperature, and humidity, for a total of six predictors. Load the file and specify the response *Evap* and the predictors *Maxat, Minat, Maxst, Minst, Maxh*, and *Minh*.

8.4.1. Check the assumption of linear predictors. Explain how you do this and your conclusions for these data.

8.4.2. Use both the inverse partial response plots and *SIR* to explore the dimension of this regression. What do you conclude? What is the evidence?

8.4.3. Verify numerically that the linear combinations h_1 and h_2 produced by *SIR* are uncorrelated. This can be done by using the "Extract Horizontal" command on the "Evaporation:SIR" plot to extract the variables on the horizontal and out-of-page axes. Next, type the command `(covariance-matrix h1 h2)` to compute the covariance matrix for h_1 and h_2, the variables on the horizontal and out-of-page axes. Verify numerically the relationship between the raw and standardized *SIR* coefficients of *maxst* in h_1.

8.4.4. Repeat the first two parts of this problem after removing the predictor *minat*. Compare the results of your analysis to those in the first two parts of this problem for the full data.

8.5. In the air quality example discussed in Section 8.4, begin by setting up the regression with *Ozone* as the response and the other three variables as predictors.

8.5.1. Fit a fifth degree polynomial to the regression of *SolR* on *Ozone*. Show that the fitted fifth degree polynomial is similar to the *lowess* fit shown in Figure 8.10a. Does the fifth degree polynomial do a better job of fitting than fitting a straight line? How do you know?

8.5.2. Draw a summary plot for these data, and describe its features. In particular, where are the 16 identified points in this plot?

CHAPTER 9

Predictor Transformations

The graphical tools presented so far work best with linear predictors. Transforming predictors to achieve linearity is generally a good way to start an analysis, but this will not necessarily produce a simple model for the data. Graphical methods can be used to tell us if linear predictors need to be transformed to achieve a simple model. In particular, we want to pick transformations of the linear predictors so that the regression function is linear in the transformed predictors. Devising a graph that will display the required transformation is the first step.

9.1 COMPONENT-PLUS-RESIDUAL PLOTS

Think of a regression problem with response y and p linear predictors x. Let x_2 be one of the predictors. Collect all remaining predictors from x into a $(p-1) \times 1$ vector x_1. We need to decide if x_2 should be transformed, and if it needs transformation, we need to find the transformation. Let $t(x_2)$ denote the appropriate transformation for x_2. We assume that a linear model is correct in the transformed scale,

$$y|x = \alpha_0 + \alpha_1^T x_1 + t(x_2) + \varepsilon \qquad (9.1)$$

The errors are assumed to be independent with mean zero and constant variance.

137

Suppose we could draw the graph $\{x_2, t(x_2)\}$. If this graph is linear, then $t(x_2) = \gamma_2 x_2$, and substituting for $t(x_2)$ in equation (9.1) leads to the linear model, and transformation is not required. If the graph is quadratic, then $t(x_2) = \gamma_2 x_2 + \gamma_{22} x_2^2$, and we can simply add the quadratic term $\gamma_{22} x_2^2$ to the usual linear model. Similarly, if t is a higher order polynomial, then we can add the corresponding polynomial terms in x_2 to the model. If x_2 is positive and $t(x_2) = \gamma_2 x_2^\lambda$, then replacing x_2 with the power transformation x_2^λ gives us a linear model. If none of these alternatives is appropriate, we can approximate t using a smoother.

When we do not know $t(x_2)$, model (9.1) has 2D structure because two linear combinations, $\alpha^T x_1$ and x_2, are needed to describe the regression function. By transforming x_2 we can possibly reduce dimensionality. Suppose we can find a new predictor z that is a good estimate of $t(x_2)$. We can then replace $t(x_2)$ by z in (9.1) and obtain the new model

$$y|x = \alpha_0 + \alpha_1^T x_1 + z + \varepsilon$$

This model has 1D structure because it depends only on the single linear combination $\alpha^T x + z$. A plot that can be used to estimate $t(x_2)$ is called a *component-plus-residual plot*.

We will use the acronym *C+R* plot to designate a component-plus-residual plot. A *C+R* plot is constructed by first getting the ols fit of the linear model,

$$y|x = \beta_0 + \beta_1^T x_1 + \beta_2 x_2 + \text{error}$$

We have written the error term on this model as a reminder that it is a *working model* and is not necessarily an adequate description of the data. From the ols fit of this model we need the ordinary residuals e and the estimate $\hat{\beta}_2$ of the coefficient of the variable x_2 that we are considering for transformation. A *C+R* plot for x_2 is the 2D plot

$$\{x_2, \hat{\beta}_2 x_2 + e\} \tag{9.2}$$

The name component-plus-residual plot comes from the quantity on the vertical axis, which is the sum of $\hat{\beta}_2 x_2$, the component of the linear model corresponding to the variable in question, and the residuals.

Both the *C+R* plot and the plot $\{x_2, t(x_2)\}$ have the same horizontal axis and differ only on the vertical axis. Let

$$\hat{t}(x_2) = \hat{\beta}_2 x_2 + e$$

In large samples we will have

$$\hat{t}(x_2) \approx t(x_2) + \varepsilon \tag{9.3}$$

where ε is the error in model (9.1). This result holds as long as $E(x_1|x_2)$ is linear in x_2, which is a weaker condition than linear predictors. In large samples a $C+R$ plot is essentially $\{x_2, t(x_2) + \varepsilon\}$. This enables us to think of a $C+R$ plot as the response plot in a simple regression problem with an unknown regression function. The unknown transformation $t(x_2)$ is the regression function for a $C+R$ plot. The key point is that we can now apply all that we know about studying simple regression functions to $C+R$ plots.

The $C+R$ plot for x_2 will display a good estimate of $t(x_2)$ as long as the errors are not too large and $E(x_1|x_2)$ is linear in x_2. When the predictors are highly nonlinearly related, the $C+R$ plot may not accurately reflect t.

9.2   PUTTING $C+R$ PLOTS TO USE

Load the file `demo-prd.lsp` from the `R-data` folder. To illustrate the use and limitations of $C+R$ plots, three items in the "TranPred" menu create artificial data sets. In each, the sample size is 100 with three predictors, $x_{11}, x_{12},$ and x_2. For all three demonstrations, the response y is generated from the model

$$y|x = 1 + x_{11} + x_{12} + x_2^{-0.67} + \varepsilon$$

so $t(x_2) = (x_2)^{-0.67}$. The ε are independent $N(0, 0.25)$. To connect with the preceding discussion, $x = (x_{11}, x_{12}, x_2)^T$ and $x_1 = (x_{11}, x_{12})^T$. Only the distribution of the predictors changes between examples.

9.2.1   Uncorrelated Predictors

Select the item "Predictors Uncorrelated" from the "TranPred" menu. In the standard regression dialog select $x_{11}, x_{12},$ and x_2 as predictors and y as the response. In this example, x_{11} and x_{12} are independent $N(0, 1)$ and x_2 has an independent uniform distribution on the interval $(0.05, 1)$. Also included in the data file are the actual values of ε, the values of $t(x_2)$, and the values of $E(y|x)$; of course in a real data problem these would not be available. Construct the plot $\{x_2, t(x_2)\}$, as shown in Figure 9.1. This plot displays the true transformation evaluated at the data points and represents the function of x_2 that we would like to recover.

Now select the item "C+R Plots–All 2D" from the regression menu. This will produce a 2D plot with an extra slide bar. The initial view in this

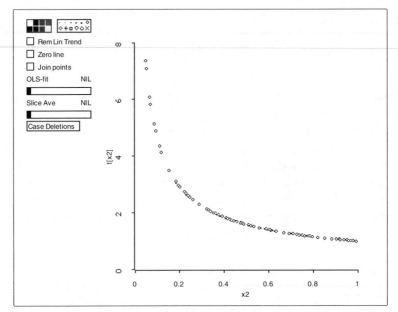

Figure 9.1. The true function of x_2 in the "Predictors Uncorrelated" demonstration.

plot is the $C+R$ plot for one of the predictors; pushing the mouse in the slide bar will cycle through all the predictors.

If you entered the predictors into the regression in the order x_{11}, x_{12}, and then x_2, the initial $C+R$ plot will be for x_{11}, as shown in Figure 9.2. This plot shows a dominant linear trend with no curvature but with nonconstant variance, since variability is larger at the right of the graph than it is at the left. A few points seem to be above the main data cloud. Mark these points with a symbol, as we have done in Figure 9.2. Since the regression function for the plot is linear, there is no evidence of the need to transform x_{11}.

The $C+R$ plot for x_{12} is qualitatively similar to the one for x_{11}, as should be expected since both were generated in the same way. The $C+R$ plot for x_2, shown in Figure 9.3, is strongly nonlinear; indeed it nearly matches the shape of Figure 9.1. This suggests transforming x_2. The points marked earlier as separate from the main point cloud are at the extremes in this plot, so those extreme points were caused by the need to transform x_2.

9.2.2 Choosing a Transformation

The next task is to estimate $t(x_2)$ using the $C+R$ plot in Figure 9.3. First, use the "OLS-fit" plot control to see if the data can be matched by fitting a polynomial. If a polynomial of degree 2 matches the data, then one

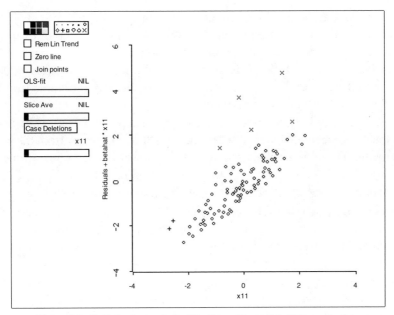

Figure 9.2. $C+R$ plot for x_{11} in the "Predictors Uncorrelated" demonstration.

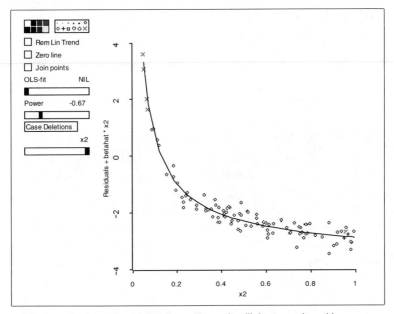

Figure 9.3. $C+R$ plot for x_2 for the "Predictors Uncorrelated" demonstration with a power smooth added.

can simply add the quadratic predictor x_2^2 to the model and refit. For this example, some exploration suggests that fitting a polynomial of degree 5 will be necessary to match the data, and even then the fit is wiggly for larger values of x_2. Using high degree polynomials in regression models is generally not very effective. Another technique may provide a simpler model and a better transformation.

Since x_2 is strictly positive, we can try fitting a power curve. Shift-click on the smoother slide bar, and select the item "Power Curve" from the pop-up menu. The slider can be used to choose a power λ in the range from -2 to 2. The curve $\{x_2, a_0 + a_1 x_2^{(\lambda)}\}$ is then superimposed on the plot, where a_0 and a_1 are the ols coefficients from the regression of the vertical axis variable on $x_2^{(\lambda)}$ for the selected value of λ. Using this slide bar, one is led to choose λ close to -0.67, reproducing the transformation that generated the data. The curve is shown in Figure 9.3.

Since the predictors are uncorrelated, one might wonder if a partial response plot of $\{x_2, y\}$ could have been used to recover $t(x_2)$. Draw the plot and see for yourself: $t(x_2)$ is nearly invisible in this plot because of the extra variation in y caused by the contributions of the other predictors.

9.2.3   Correlated Predictors

The item "Predictors Linearly Related" in the "TranPred" menu repeats this last example, except now $x_{1j} = 3x_2 + 0.57N(0, 1)$, for $j = 1, 2$. As before, x_2 is uniformly distributed on the interval $(0.05, 1)$. The only change from the last example is that x_1 and x_2 are linearly related. This means that $E(x_{11}|x_2)$ and $E(x_{12}|x_2)$ are linear in x_2, and these are precisely the conditions that allow $C+R$ plots to be effective. Analysis of this example is left to Exercise 9.1.

9.2.4 $C+R$ Plots with Nonlinear Predictors

For the $C+R$ plot to fail, we need at least one of the regression functions $E(x_{11}|x_2)$ or $E(x_{12}|x_2)$ to be nonlinear. As an example of this, we choose x_2 as before, but now we set $x_{11} = |x_2 - .5|$ and $x_{12} = N(0, 1)$. These predictors can be obtained by selecting "Predictors Nonlinearly Related" from the demonstration menu. After setting up the regression, select the "C+R Plots–All 2D" item. The initial frame of this plot is shown in Figure 9.4.

The $C+R$ plot for x_{11} is somewhat curved, suggesting that x_{11} may need to be transformed. Removing the linear trend in the plot and then using a smoother may make the curvature more apparent. Change to the $C+R$ plot for x_2, as shown in Figure 9.5. Curvature is more apparent, but the

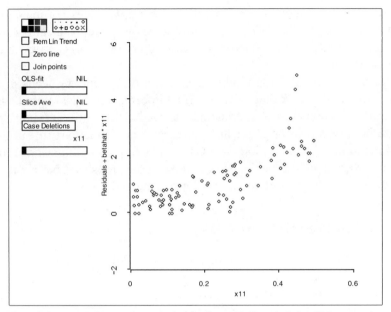

Figure 9.4. $C+R$ plot for x_{11} for the in the "Predictors Nonlinearly Related" demonstration.

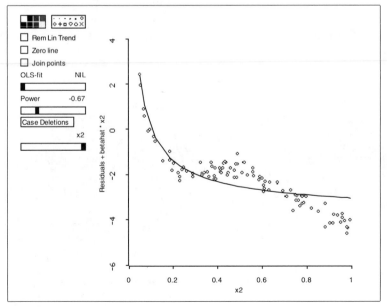

Figure 9.5. $C+R$ plot for x_2 in the "Predictors Nonlinearly Related" demonstration.

correct transformation, superimposed on Figure 9.5, does not match the points in the plot. The $C+R$ plot for x_{12}, not shown here, is linear because we constructed x_{12} to be independent of the other predictors and x_2 enters the model linearly.

Because $E(x_{11}|x_2)$ is a nonlinear function of x_2, and x_2 needs to be transformed, the $C+R$ plots give incorrect information. One might be led incorrectly to transform x_{11}. Although the need to transform x_2 is apparent from the curvature in the graph, the appropriate transformation to use cannot be determined from the plot. The $C+R$ plot can provide a powerful tool for uncovering appropriate transformations of the predictors, but not if the predictors themselves are highly nonlinearly related.

9.3   EXAMPLE: ETHANOL DATA

Return to the ethanol data described in Section 7.3 and last discussed in Section 8.3. Load the data file `ethanol.lsp` from the `R-data` folder and specify *NOx* as the response and *C* and *E* as the predictors.

Since the predictors are nearly linearly related (see Section 8.3), $C+R$ plots can be used to help choose transformations. Select "C+R Plots–All 2D" from the "Ethanol" menu. The $C+R$ plot for *E* is shown in Figure 9.6. From this plot, $t(E)$ is surely not a linear function of *E*. Might a quadratic

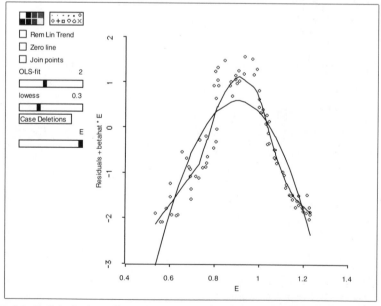

Figure 9.6. $C+R$ plot for the predictor E in the ethanol data with *lowess* and quadratic fits superimposed.

be appropriate? Use the "OLS-fit" slide bar to superimpose a quadratic on the plot, as shown in Figure 9.6. A quadratic doesn't do a very good job of describing $t(E)$.

A power function won't work here because the $t(E)$ is not monotonic, but a smoother can be used to estimate $t(E)$. A *lowess* smooth is shown in Figure 9.6. Let's agree that the *lowess* smooth is a reasonable estimate of $t(E)$ and call the smoothed transformation $t_{est}(E)$. We don't have an analytic form for $t_{est}(E)$, but its values at the data points, $t_{est}(E_i)$, can be obtained by using the "Extract Curve" plot control, obtained by shift-clicking on the *lowess* slide bar. Name the extracted variable *TE*. The transformed variable *TE* can be used in plots or in new models, just like any other variable. To see the transformed variable, construct the plot {E, TE}. The points are just values along the *lowess* smooth in Figure 9.6.

This is the most general method of estimating t from a *C+R* plot. We could now refit the model with *NOx* as the response and *C* and *TE* as the predictors and continue the analysis.

9.4 TRANSFORMING TWO PREDICTORS

The same ideas can be used to transform two predictors simultaneously. Let x_2 and x_3 denote the predictors that may require transformation and collect the remaining predictors from x into the $(p-2)$-dimensional vector x_1. The required transformation will be denoted by $t(x_2, x_3)$.

9.4.1 Models for Transforming Two Predictors

When considering the possibility of transforming two predictors, we need to distinguish between two general types of transformations: *additive* and *nonadditive*. An additive transformation is of the form $t(x_2, x_3) = t_2(x_2) + t_3(x_3)$ so the predictors are transformed individually and then the transformed values are added to get the joint transformation t. A nonadditive transformation is any transformation that is not additive. For example, $t(x_2, x_3) = x_2 x_3$ is a nonadditive transformation while $t(x_2, x_3) = x_2^2 + \log(x_3)$ is an additive transformation.

The model we assume for transforming two predictors is a straightforward extension of the model for transforming a single predictor:

$$y|x = \alpha_0 + \alpha_1^T x_1 + t(x_2, x_3) + \varepsilon$$

A *C+R* plot for x_2 and x_3 is constructed by getting the ols fit of the working model

$$y|x = \beta_0 + \beta_1^T x_1 + \beta_2 x_2 + \beta_3 x_3 + \text{error}$$

and then drawing the 3D plot

$$\{x_2, \hat{\beta}_2 x_2 + \hat{\beta}_3 x_3 + e, x_3\} \tag{9.4}$$

As in the case of transforming a single predictor, this plot will look like the plot $\{x_2, t(x_2, x_3)+\varepsilon, x_3\}$ in large samples as long as $E(x_1|x_2, x_3)$ is a linear function of x_2 and x_3. The distinction between additive and nonadditive transformations is not important for constructing the plot but may be useful for interpretation and for constructing transformations.

9.4.2   Example: Plant Height

One use for examining two predictors at once is to check for *spatial trends*. For example, agricultural field trials are often done on a rectangular field that is divided into a number of plots. Even if all the plots were treated identically, the expected responses may differ. Often these differences are systematic; for example, the expected response may decrease from north to south. The 3D $C+R$ plot provides a method of studying this sort of variation after the experiment is done.

Consider an experiment done in the spring of 1951 to investigate the effects of varying dosages of cathode rays on the growth of tobacco seeds. Seven levels of dosage were used, one of which was a control of dose zero. The experimental area was divided into 56 plots laid out in a grid with eight rows and seven columns. As is common in an experiment of this type, the experimenters believed that spatial trends might be possible, so they laid out the experiment as a *randomized complete block design* using the rows as blocks. This means that within each row, each of the seven treatments appears exactly once, allocated to plots within the row at random. Because of the blocking, any systematic effects due to rows should affect each treatment equally, and they should therefore be eliminated from comparisons between treatments. Any spatial trends due to *columns*, however, have not been eliminated from comparisons.

For each of the field plots we know the row number *Row*, the column number *Col*, the treatment number *Trt*, and the response *Ht*, the total height in centimeters of twenty plants.

If all plots are treated alike, we can imagine a *response surface* over the experimental area that reflects the spatial trends. We investigate this surface using a 3D $C+R$ plot. The *Row* and *Col* indices of a plot are coordinates of points on the surface. The original source of the paper did not give the quantitative values for the treatment levels so we will regard *Trt* as a qualitative *factor*, using indicator variables to represent the individual treatments. Specifically, let the indicator variable $v_j = 1$ if the jth treatment level is applied on the plot and 0 otherwise. We get the following general model for the experiment:

$$Ht = \alpha_0 + \alpha_1 v_1 + \ldots + \alpha_6 v_6 + t(Row, Col) + \varepsilon$$

where $t(Row, Col)$ is the value of the response surface for a particular value of *Row* and *Col*. Although the experiment had seven treatment levels, only six indicator variables are needed because the seventh is redundant in a model that includes an intercept. To construct a 3D *C+R* plot, we need to fit the working model,

$$Ht = \beta_0 + \beta_1 v_1 + \cdots + \beta_6 v_6 + \beta_7 Row + \beta_8 Col + \text{error}$$

The file `plant-ht.lsp` in the R-data folder contains the data for this experiment. Load the file, and when the regression dialog appears, push the "Factors. . . " button to create the indicator variables associated with the treatments. In the factors dialog double-click on "Trt" to move it from the left list to the right and then click "OK." This creates the indicator variables associated with the treatments. They are represented collectively by the single symbol {*F*}*Trt* in the new regression dialog. The "{*F*}" reminds us that this is a factor. Adding a factor to a model adds the set of indicator variables. The names of the indicators are of the form *Trt*[4], which would be the indicator variable that has the value 1 whenever *Trt* = 4 and is zero otherwise. The function cut described in Appendix A.7.2 can be useful in defining factors.

We are now ready to specify the working model: Designate *Row*, *Col*, and {*F*}*Trt* as predictors and *Ht* as the response. This is not the usual model for a randomized complete block design, which would use {*F*}*Trt* and {*F*}*Row* as predictors, where {*F*}*Row* is a factor for rows to represent the blocking effects. *Col* would not appear in the usual randomized complete block analysis. The analysis here could be used in any problem in which the units have coordinates in the plane, regardless of the experimental design.

Part of the point of this example is to introduce how factors are created in the *R-code*. Now we can get back to *C+R* plots. From the regression menu called "Plant-Ht" select the item "C+R plot–3D" and in the next dialog move *Col* and *Row* to the right selection box and then click "OK." The plot on the screen is the 3D *C+R* plot for *Col* and *Row*. It should give us information on the underlying surface $t(Row, Col)$.

Use the "Pitch" control to rotate the 3D *C+R* plot to the 2D view {*Row, Col*}. The points fall on a regular 8×7 grid corresponding to the

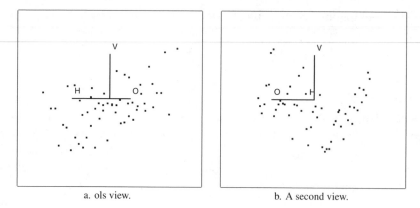

a. ols view.   b. A second view.

Figure 9.7. Two views of the 3D $C+R$ plot for the plant-height data.

field layout. Now return the plot to the "Home" position and rotate the plot by using one of the "Yaw" buttons to gain a feeling for the function $t(Row, Col)$. What do you see? Is the structure of the plot 1D or 2D? Two 2D views of the plot are shown in Figure 9.7. Substantial spatial differences across the field are evident. The plot seems to be composed of a linear trend, visible in the ols view in Figure 9.7a, and a quadratic trend, shown in Figure 9.7b. Because two views are required, the structure must be 2D. This is useful information, particularly for future experimental designs in the same area, but it would be a help in the analysis of this particular experiment if we could characterize the surface more specifically.

Recall the ols view and then print the screen coordinates:

```
Linear Combinations on screen axes in current rotating plot.
Horizontal: 0.481996  + 0.160968 H + 0 V + -0.250193 O
Vertical:   0.0133144 + 0 H + 0.00252048 V + 0 O
```

Since both *Col* and *Row* contribute about equally to the horizontal screen variable, the strongest linear trend runs diagonally across the field and thus does not align with blocks, which were rows. Next, remove the linear trend and view the resulting plot while rotating. Is the detrended plot 1D or 2D? One view of the detrended plot is shown in Figure 9.8. The detrended plot seems to be dominated by 1D structure, although relatively minor deviations are apparent as well. This suggests using the nonadditive form

$$t(Row, Col) = b_1 Row + b_2 Col + t_1(a_1 Row + a_2 Col)$$

where $b_1 Row + b_2 Col$ represents the combination giving the strongest linear trend and $t_1(a_1 Row + a_2 Col)$ is the 1D structure in the detrended plot. The strongest 2D view of the detrended $C+R$ plot seems to be the view with *Col*

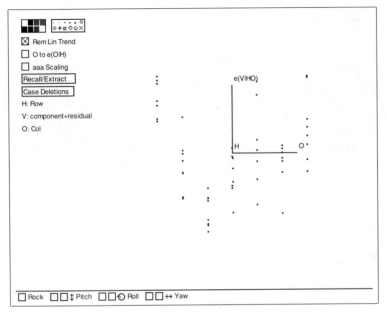

Figure 9.8. Detrended 3D $C+R$ plot for the plant-height data.

on the horizontal screen axis, which implies that $a_1 = 0$. This simplifies the nonadditive transformation to an additive form

$$t(Row, Col) = b_1 Row + b_2 Col + t_2(Col)$$

The punch line for our graphical analysis of the 3D $C+R$ plot is that we can reduce the problem to a transformation of Col alone, leading to the simplified model

$$Ht = \alpha_0 + \alpha_1 v_1 + \cdots + \alpha_6 v_6 + \alpha_7 Row + t(Col) + \varepsilon \qquad (9.5)$$

To determine the transformation to use in (9.5) we can use the 2D $C+R$ plot for Col, as shown in Figure 9.9. Neither the quadratic nor the *lowess* fit shown on the plot provides a fully satisfactory approximation to t. The quadratic fails to capture the points at $Col = 3$. Because of the few discrete values of Col, the *lowess* smooth just connects the averages of each group of points. If the primary goal is reducing variation to allow more powerful treatment comparisons, then using the quadratic may be sufficient. At the other extreme, Col could be included as a factor, which is essentially the solution suggested by the *lowess* curve. The form of t in this case is

$$t(Col) = \gamma_1 c_1 + \cdots + \gamma_6 c_6 \qquad (9.6)$$

where the c_k are the indicator variables for columns.

The analysis of the plant-height data continues in Exercise 9.4.

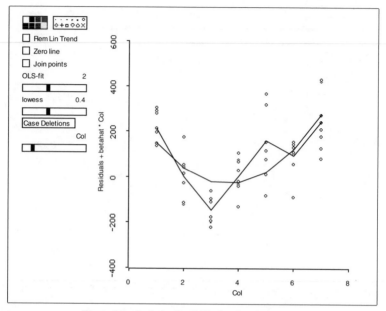

Figure 9.9. *C+R* plot for *Col* in the plant-height data.

9.5 TRANSFORMING MANY PREDICTORS

A restriction so far in this chapter is that only one or two of the predictors might need to be transformed, while the others enter the model linearly. When more than two of the predictors may need to be transformed, an adaptation of the basic methodology presented here may yield useful results.

After fitting the linear working model, inspect the 2D *C+R* plots for convincing nonlinear trends. If none are found, there is no evidence that any predictors should be transformed. Otherwise, identify the predictor with the smallest variation about a nonlinear trend and then choose a transformation of that predictor. In a generic model, this means first obtaining an estimate $t_{est}(x_2)$ of the transformation as a curve on a *C+R* plot and then extracting the estimate by using the "Extract Curve" plot control. The estimate could be a polynomial, a power, or a smooth.

Next, subtract $t_{est}(x_2)$ from the response and construct the new working model:

$$y - t_{est}(x_2) = \beta_0 + \beta_1^T x_1 + \text{error} \tag{9.7}$$

This is a linear model with response $y - t_{est}(x_2)$ and linear predictors x_1, so we can use *C+R* plots to explore the need to transform the components of x_1.

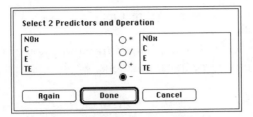

Figure 9.10. The interactions dialog.

In the ethanol data, we can examine the need to transform C, given that E has been replaced by TE. Fit a working model with $NOx - TE$ as the response and single predictor C. To obtain this response variable, assuming that TE has already been extracted, select the item "New Model..." from the regression menu. This will give you a standard regression dialog. To create the new variable, $NOx - TE$, push the "Interactions..." button in the dialog. You will then get a dialog similar to Figure 9.10. Select the two variable names and the operation "$-$" as shown and then push "Done." This will return you to the main regression dialog; choose $NOx - TE$ as the response and C as the predictor.

With a single predictor, the $C + R$ plot is equivalent to a standard plot of the predictor versus the response, as shown in Figure 9.11. This plot contains

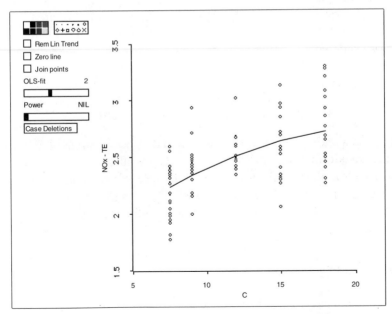

Figure 9.11. Plot to choose a transformation of C in the ethanol data.

a slight hint of a nonlinear trend that may be modeled by a quadratic, as shown on the plot.

Suppose we decide that a quadratic transformation of C is worth considering. Combining the transformations, we then have the new working model

$$NOx = \beta_0 + \beta_1 C + \beta_{11} C^2 + \beta_2 TE + \text{error} \qquad (9.8)$$

There is no guarantee that this model gives a good description of the data, although models constructed in this way are often adequate. One potential problem is that transforming predictors individually permits only additive transformations and does not allow for interaction terms like $E \times C$. In any event, the model should always be checked by using residual plots, as to be discussed in Chapter 11.

The ethanol data have only two predictors, but the same ideas apply for any number of predictors. After transforming q predictors, $x_{21}, \ldots, x_{2q}$, the working model is of the form

$$y^* = \beta_0 + \beta_1^T x_1 + \beta_2 x_2 + \text{error}$$

where

$$y^* = y - t_{\text{est}:1}(x_{21}) - \cdots - t_{\text{est}:q}(x_{2q})$$

y^* is the original response minus all of the extracted transformation curves to date and x_2 is the next variable to be transformed. The transformation of x_2 is determined as usual, and the extracted transformation curve is then subtracted from y^*. The process continues until all relevant predictors have been considered.

As a final check on the transformations, inspect the C+R plots for the working model

$$y^* = \beta_0 + \beta_1 x_1 + \cdots + \beta_p x_p + \text{error}$$

where the working response y^* is the original response minus all of the extracted transformation curves and $x_1, \ldots, x_p$ are all the original predictors. If all the C+R plots are linear, then no further transformations are required. If some are nonlinear, then further transformations may be useful. Details of this iterative fitting method are available in the references given in the next section.

9.6   COMPLEMENTS

The component-plus-residual plot was first suggested by Ezekiel (1924); Wood (1973) is responsible for the name. The C+R plots are sometimes called *partial residual plots*. As described here, the $C+R$ plots start with the ols fit of a working model, but the role of ols is not crucial, and it could be replaced by many other estimation methods. The treatment of $C+R$ plots in this chapter is based on the recent work of Cook (1993), who describes a larger class of plots called *CERES* plots that do not require the assumption of linear predictors to determine appropriate transformations.

A formal iterative process of model building that uses $C+R$ plots is called *generalized additive modeling* and was developed by Hastie and Tibshirani (1990). The general idea, which originated with Ezekiel (1924), is to cycle through all relevant predictors until the transformations no longer change. One cycle through the relevant predictors is usually sufficient with linear predictors.

The plant-height data are from Federer and Schlottfeldt (1954). The data for Exercise 9.5 are from Federer (1955).

EXERCISES

9.1. *9.1.1.* Analyze the data from the item "Predictors Linearly Related" in the "TranPred" menu following the steps in the text for the "Predictors Uncorrelated" demonstration in Section 9.2.1. Compare the results of these two demonstrations.

9.1.2. What is the structural dimension of the model used to generate these two demonstrations? Without transforming x_2, analyze the data for the "Predictors Linearly Related" demonstration using inverse partial response plots and *SIR*. Do these methods find the structure?

9.2. A $C+R$ plot, defined by (9.2), is a plot of $\{x_2, \hat{\beta}_2 x_{i2} + e_i\}$. Suppose we use ols to fit the equation

$$\hat{t}(x_2) = \gamma_0 + \gamma x_2 + \text{error}$$

Verify that $\hat{\gamma}_0 = 0$ and that $\hat{\gamma} = \hat{\beta}_2$, where $\hat{\beta}_2$ is the ols estimate of β_2 in the working model

$$y|x = \beta_0 + \beta_1^T x_1 + \beta_2 x_2 + \text{error}$$

that is used as the basis for constructing $C+R$ plots. This can be done by algebraically manipulating the usual formula for the slope in a simple linear regression or by using a numerical example to illustrate that $\hat{\gamma} = \hat{\beta}_2$ apart from rounding error.

Using these results, explain why a detrended $C+R$ obtained by using the "Rem Lin Trend" plot control, is just the plot $\{x_2, e\}$.

9.3. In the Big Mac data in the file `big-mac.lsp` in the R-data folder, investigate the need to transform the predictors in the working model

$$\log(BigMac) = \beta_0 + \beta_1 TeachTax + \beta_2 TeachSal + \beta_3 Service + \text{error}$$

Inspect the $C+R$ plots for the three predictors in the working model. Which predictor x_2 is the most likely to need a transformation? Determine an appropriate transformation by superimposing a curve on the plot. The curve could be based on fitting a polynomial using the ols slide bar, or it could be a power curve or a smoother. Follow the steps in Section 9.5 to see if another predictor needs to be transformed. If so, repeat the entire procedure to see if the remaining predictor requires transformation. State your final transformations.

9.4. Complete the analysis of the plant-height data in Section 9.4.2 by comparing conclusions on treatment effects under three versions of model (9.5). In the first, set $t(C) = \alpha_8 C$. In the second, set $t(C) = \alpha_8 C + \alpha_9 C^2$. Finally, set $t(C)$ to be as given in equation (9.6).

9.5. The file `rubber.lsp` in the R-data folder contains the results of an experiment to compare the rubber yield of seven varieties of guayule. The experimental area consisted of 35 plots arranged in a 5×7 grid, the rows of the grid forming five randomized complete blocks. The response for this problem is the total grams $P_1 + P_2$ of rubber for the two selected plants on each plot.

Aside from needing to add the variables P_1 and P_2 to obtain the response, the structure of these data is the same as those in Section 9.4.2. Conduct a graphical analysis of these data following the rationale and general steps of Section 9.4.2. Were the blocks selected to be in the best direction?

CHAPTER 10

Response Transformations

The predictor transformations discussed in Chapter 9 are used to simplify a problem by reducing dimensionality. Response transformations are used to linearize the regression function in problems with 1D structure and to get a constant variance function.

10.1 REGRESSIONS WITH ONE PREDICTOR

Suppose we have a simple regression problem with response y and single predictor v. If the response plot $\{v, y\}$ is clearly curved, then we will know that the regression function

$$E(y|v) = f(v)$$

is a nonlinear function f of v. Sometimes a nonlinear relationship can be turned into a linear one by a suitable *monotonic* transformation $t(y)$ of y. The regression model is then linear in the transformed scale,

$$t(y)|v = \beta_0 + \beta v + \varepsilon \qquad (10.1)$$

Analyzing a model that is linear in v on a transformed scale is often much easier than analyzing a model that is nonlinear in v in the original scale.

Consider the response plot $\{v, y\}$ shown in Figure 10.1a. You can reproduce this figure by loading the file resptran.lsp from the R-data

155

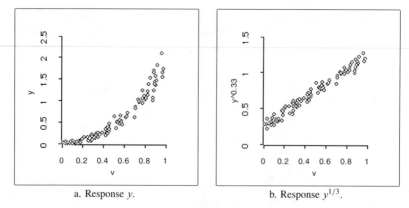

a. Response y.   b. Response $y^{1/3}$.

Figure 10.1. The regression function for the summary plot (a) is curved. In (b), y has been replaced by $y^{1/3}$, and the regression function is linear. The variance function is also different in the two plots.

folder. Since this plot is nonlinear, we should not use a linear regression function to model the data. Figure 10.1b is obtained from Figure 10.1a by replacing y with $t(y) = y^{1/3}$. No curvature is apparent in this plot, and so we could proceed by fitting the linear model (10.1) with $t(y) = y^{1/3}$. The plot in Figure 10.1b provides a visualization of the regression function $E(t(y)|v)$. Transforming the response in this example has linearized the regression function and changed the variance function. In Figure 10.1a, the variance appears to increase with v, while in Figure 10.1b the variance function appears to be constant. The transformation thus achieved a linear regression function and a constant variance function. Achieving both of these goals by a single transformation is often possible.

In the example we knew the appropriate transformation $t(y) = y^{1/3}$, but we will not normally have such information. How can we determine a suitable response transformation graphically in real problems? The plot in Figure 10.1a does not help because the regression function is $E(y|v) = f(v)$, not $t(y)$. However, the *inverse response plot* $\{y, v\}$ can help us choose a suitable transformation when the untransformed regression function $E(y|v)$ is monotonic, as in Figure 10.1a. If $E(y|v)$ is not monotonic, then a response transformation may not be appropriate.

An inverse response plot is useful for choosing a transformation $t(y)$ when $E(y|v)$ is monotonic because then

$$E(v|y) \approx t(y) \tag{10.2}$$

This equation tells us that the regression function for the inverse response plot $\{y, v\}$ is approximately the required transformation. The transfor-

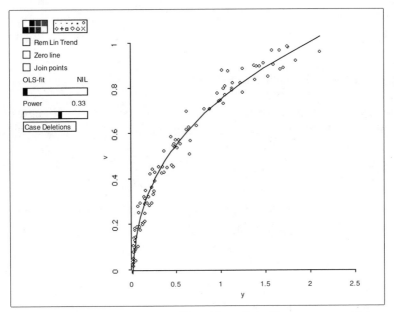

Figure 10.2. Inverse response plot corresponding to Figure 10.1a. The power curve added suggests transforming y to $y^{1/3}$.

mation itself can be estimated by fitting a curve to the plot, much like transformations of predictors were selected by fitting curves to $C+R$ plots in the last chapter. The most useful method is to use a power transformation, but other choices such as fitting a polynomial or a smoother can be used. Once an appropriate estimate is determined, the transformed values can be extracted and then used as the response in further analysis. If a power curve is used giving a power λ, then the transformed response $y^{(\lambda)}$ can be created from the "Transform. . ." dialog.

For the example in Figure 10.1a, the inverse response plot $\{y, v\}$ is shown in Figure 10.2. The plot contains a power curve obtained by using the "Power Curve" option from the smoother slide bar, with power 0.33. Since the curve matches the data, the cube-root transformation is suggested. Figure 10.1b shows the response plot in the cube-root scale.

Equation (10.2) will hold exactly when the predictor v and the errors ε in (10.1) have the same properties as a pair of linear predictors. Since the errors are unknown, we cannot draw a plot to check this property. Equation (10.2) will be a good approximation if $t(y)$ is quite a bit bigger than ε so the signal dominates the noise. The bottom line is this: If the plot $\{y, v\}$ shows a well-determined curve, then (10.2) is a good approximation and the curve can be used to guide the selection of a transformation. If the plot $\{y, v\}$

does not show a well-determined curve, then a response transformation will probably not aid the analysis by linearizing the regression function.

10.2   MANY LINEAR PREDICTORS

The ideas discussed in the previous section can be extended to regression problems with many predictors and 1D structure. Under this assumption the regression function is

$$E(y|x) = f(\beta^T x) \tag{10.3}$$

where f is unknown. As with one predictor, we want a monotonic transformation $t(y)$ that gives a linear regression function

$$E(t(y)|x) = \beta_0 + \beta^T x \tag{10.4}$$

If we knew $c\beta$ for some nonzero constant c, then we could set $v = c\beta^T x$ and use the inverse plot $\{y, v\}$ as described in the last section. Since the parameter β is not known, the plot $\{y, c\beta^T x\}$ cannot be drawn. We can make progress only if we can get an estimate of $c\beta$. Can this be done when we don't know f? From the 1D estimation result in Section 7.4.1, the ols coefficients $\hat{b}$ from the regression of y on x provide an estimate of $c\beta$ without knowing f as long as we have linear predictors. This means that, given linear predictors, the plot $\{y, \hat{b}^T x\}$ or equivalently the plot $\{y, \hat{y}\}$ may show an appropriate transformation t, up to the random errors in the data and unimportant multiplication by a constant. We will call the plot $\{y, \hat{y}\}$ an *inverse fitted-value plot*.

As an example, we use data from a designed experiment on the strength of worsted yarn. The response variable y is the number of loading cycles to failure. There are three predictors: the length of test specimen x_1, the amplitude of loading cycle x_2, and the load x_3. The response was measured at all possible combinations of three settings for each predictor, resulting in $3^3 = 27$ observations. Load the file wool.lsp from the R-data folder, and specify y as the response and x_1, x_2, and x_3 as predictors. The fitted values $\hat{y}$ from this regression will be used to study the need for a transformation of the response.

Draw the plot $\{y, \hat{y}\}$ as shown in Figure 10.3. If the regression function for Figure 10.3 were a straight line, no transformation would be indicated. Since the plot is curved and the regression relationship is monotonic, a transformation is needed. All of this is under the assumption of 1D structure.

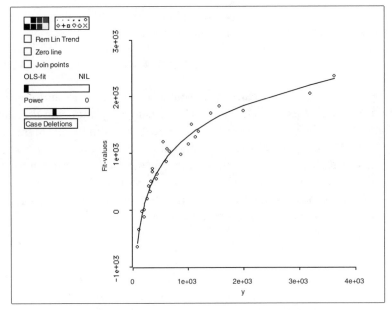

Figure 10.3. Inverse fitted value plot for the wool data.

The next question is the choice of the transformation, and this can be determined visually by adding a fitted curve to the plot. Except for the monotonicity requirement, this is exactly as was done for choosing predictor transformations in Section 9.2.2. Since the response is strictly positive, we can try power curves. Shift-click on the "Slice Ave" slide bar, and from the pop-up menu select "Power Curve." Shown in Figure 10.3 is the curve for $\log(y)$, which seems to match the data quite well, suggesting that $\log(y)$ is appropriate.

10.3   NUMERICAL CHOICE OF TRANSFORMATION

The inverse fitted-value plot for choosing a response transformation can be augmented by a numerical method, often called the *Box–Cox* method. The two methods are complementary and will often give the same result. The inverse fitted-value plot requires linear predictors and 1D structure, while the Box–Cox method requires 1D structure, specification of a family of transformations indexed by one parameter, and approximate normality of the errors with the transformed response. With the numerical procedure we will obtain both a point estimate and an interval estimate of a transformation parameter.

We now assume that there is a scaled power transformation $y^{(\lambda)}$ of the response so that the linear model holds in the transformed scale,

$$y^{(\lambda)}|x = \beta_0 + \beta^T x + \varepsilon \qquad (10.5)$$

where the errors are normally distributed with mean zero and constant variance. Given this model, we might estimate λ by choosing the value that minimizes the residual sum of squares $\text{RSS}(\lambda)$ from the ols regression of $y^{(\lambda)}$ on x. This general idea is right, but the details are wrong because the units of $\text{RSS}(\lambda)$ are different for every value of λ. Consequently, we can't compare values of $\text{RSS}(\lambda)$ for different values of λ. A way out of this problem is to adjust the transformation so that the units are always the same. Define a *modified power transformation* $z(\lambda)$ by

$$z(\lambda) = y^{(\lambda)} \text{gm}(y)^{1-\lambda}$$

where $\text{gm}(y)$ is the geometric mean of the observed values of y. The estimate of λ minimizes $\text{RSS}_z(\lambda)$, the residual sum of squares from the regression of $z(\lambda)$ on x.

Finding the value of λ that minimizes $\text{RSS}_z(\lambda)$ can be done graphically using *confidence curves* as displayed in the plot

$$\{[n(\log(\text{RSS}_z(\lambda)) - \log(\text{RSS}_z(\hat{\lambda})))]^{1/2}, \lambda\}$$

For the wool data, the confidence curves are given in Figure 10.4. The value $\hat{\lambda}$ that minimizes $\text{RSS}_z(\lambda)$ is the point where the curves meet the vertical axis, which is $\hat{\lambda} = -0.05$ for the wool data. The confidence curves can be used to get an interval estimate for λ. The horizontal axis of Figure 10.4 is labelled `z-value`. Values on this axis correspond to the standard normal distribution. For example, the interval between the two curves at 1.96 on the horizontal axis, which is approximately -0.18 to $+0.06$, is a 95% confidence interval for λ because the area under a standard normal curve between -1.96 and $+1.96$ is 95%. As with the graphical procedure in the last section, the log transformation is suggested because $\hat{\lambda}$ is close to zero.

To draw Figure 10.4, select the item "Choose Response Transform" from the regression menu. You will get a dialog to choose the transformation family and whether or not a constant is to be added to the response before transformation. The defaults are appropriate here, so just push the "OK" button. The other choices are discussed in Section 10.5. This plot requires a lengthy calculation.

We now have two methods for choosing a response transformation, the inverse fitted-value plot described in Section 10.2 and the Box–Cox method.

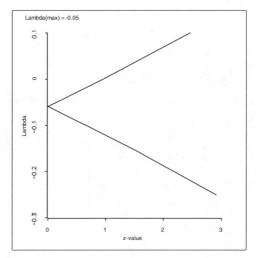

Figure 10.4. Confidence curves for choosing a transformation in the wool data.

For the wool data, these two gave the same transformation, but they need not always agree. The inverse fitted-value plot chooses transformations to linearize the regression function, while the Box–Cox method tries to make the residuals in the transformed scale as close to normally distributed as possible. For example, suppose that $E(y|x)$ were independent of x but with errors that have a skewed distribution. The inverse fitted-value plot will suggest no transformation, while the Box–Cox method will choose a transformation to make y more nearly normally distributed.

10.4   EXAMPLE: MUSSELS' MUSCLES

In this section we consider an example to bring together some of the ideas in the past few chapters. The data come from a study of Horse mussels sampled from the Marlborough Sounds, which are located off the Northeast coast of New Zealand's South Island. The response variable is the muscle mass M, the edible portion of the mussel, in grams. There are four predictors all relating to characteristics of mussel shells: shell width W, height H, and length L each in millimeters and shell mass S in grams. The goal of the example is to develop an understanding of how the distribution of muscle mass depends on the four predictors. We expect that the regression function $E(M|L, W, H, S)$ increases with the values of the predictors. Quantifying just how such increase takes place is part of this study. The data are contained in the file `mussels.lsp` in the `R-data` folder.

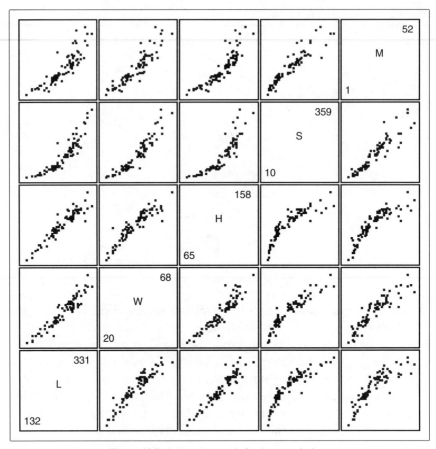

Figure 10.5. Scatterplot matrix for the mussels data.

We begin by constructing a scatterplot matrix of all the variables, as shown in Figure 10.5. The partial response plots in the top row of the scatterplot matrix show the expected increasing relationship. The partial response plots for length, height, and width show similar curvature, while the plot for shell mass appears linear. All partial response plots suggest that the variability of muscle mass increases with the predictors. The inverse partial response plots in the last column of the scatterplot matrix exhibit the same general relationships as the partial response plots. If the predictors were linear, this would be enough to suggest that the structural dimension of the problem is probably 2, as two linear combinations of the predictors are needed. (Why is this so?) The pairwise relationships between length, width, and height all appear linear, but the pairwise relationships involving shell mass are nonlinear. Thus the scatterplot matrix does not sustain the

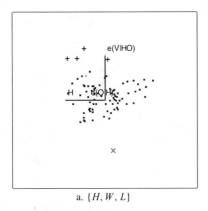
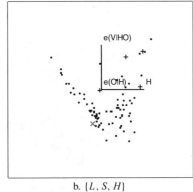

a. $\{H, W, L\}$ b. $\{L, S, H\}$

Figure 10.6. Most curved 2D views of two 3D plots of predictors in the mussels data. In each plot the "Rem Lin Trend" and the "O to e(O|H)" buttons have been pushed.

assumption of linear predictors, and using the inverse response plots or *SIR* to infer about structural dimension may be premature.

To get a closer look at the linearity of the predictors, inspect the rotating plot $\{H, W, L\}$ while using the "Rem Lin Trend" and "O to e(O|H)" options. Were these three predictors all linearly related, then this plot should approximate a spherical swarm of points. The 2D view of this plot with the strongest pattern we could find visually is shown in Figure 10.6a. Some curvature is apparent but not enough to question seriously the applicability of methods based on linear predictors. One relatively large and negative outlying point is visible in the figure. The four points with relatively large, positive values of $e(V|HO)$ are probably consequences of nonconstant variance; to check this, you might select these points and note their location in the partial response plots in the scatterplot matrix.

In contrast to this 3D plot, a relatively curved view of the 3D plot $\{L, S, H\}$ is shown in Figure 10.6b. The curvature here is quite strong and certainly enough to rule out the assumption of linear predictors.

Using the transformation slide bars on the scatterplot matrix, we examine power transformations of S to make its transformed value linearly related to the other predictors. The cube root of S appears to achieve linearity. The adjacent transformations, log and square root, are ruled out because the corresponding plots are curved.

Since only one of the predictors needs to be linearized, we can use a complementary approach. Of interest is to make the regression function $E(S|L, W, H)$ linear. We can do this by examining the regression problem with S as a response variable and L, W, and H as predictors. The methodology of Section 10.3 can be used to choose a transformation of S that gives

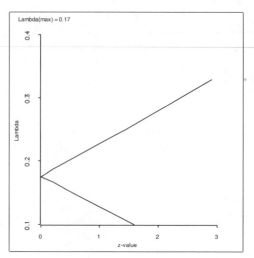

Figure 10.7. Confidence curves for a power transformation of S in the regression on H, W, and L for the mussels data.

approximate linearity. Use the "New Model. . . " item from the "Mussels" menu to set up a regression of S on the other three predictors (H, W, L). From the resulting "Mussels1" menu select "Choose Response Transform" and use the default options in the dialog. The confidence curves for the power of S are shown in Figure 10.7. They indicate that the 0.2 power is a somewhat better transformation than the cube root. Inspect the 3D plot $\{L, S^{0.2}, H\}$. To get this plot, select "New Model. . . " from the regression menu, and in the regression dialog, push the "Transform. . . " button. In the transformation dialog, select S, type ".2" in the "Power" box, and then push "Done." When you return to the regression dialog, you can either create a new model or just push the "Cancel" button. In either case, the new variable $S^{0.2}$ will appear whenever you choose the "Plot of. . . " or "New Model. . . " items. Did the power transformation successfully remove the curvature of Figure 10.6b?

To continue the analysis, construct a scatterplot matrix of the response and the new set of predictors, as shown in Figure 10.8. The predictors now appear linear, and the inverse partial response plots all have the same shape, suggesting 1D structure. Results from *SIR* support this conclusion. This is a very nice outcome since 2D structure was indicated for the original problem in the scatterplot matrix of Figure 10.5. Response transformations may simplify the analysis further by linearizing the regression function.

Set up the regression with M as the response and L, W, H, and $S^{0.2}$ as the predictors, and then draw the inverse fitted-value plot $\{M, \hat{M}\}$, as shown in Figure 10.9. This plot supports the conclusion that λ values

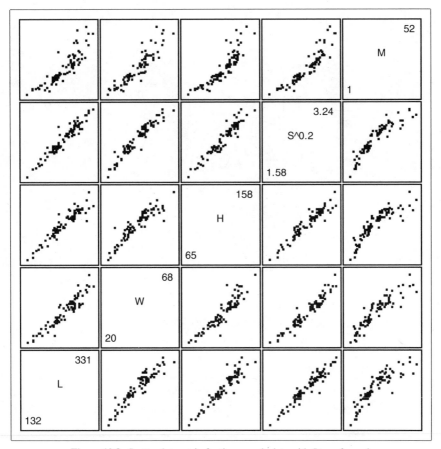

Figure 10.8. Scatterplot matrix for the mussels data with S transformed.

of 0, 0.25, or 0.33 may work, as all of these power curves fit about equally well.

The Box–Cox procedure indicates using the cube-root transformation. However, this may be due in part to case 47, which is in the lower left of Figure 10.9 and is relatively far from the fitted power curve when the log transformation is used.

We have now arrived at the following model:

$$M^{1/3}|x = \beta_0 + \beta_1 L + \beta_2 W + \beta_3 H + \beta_4 S^{0.2} + \varepsilon \qquad (10.6)$$

The next step in the analysis is to apply various graphical diagnostics as checks on the model to see if refinement is necessary. We begin our discussion of model checking in the next chapter. In the meantime, a $C+R$

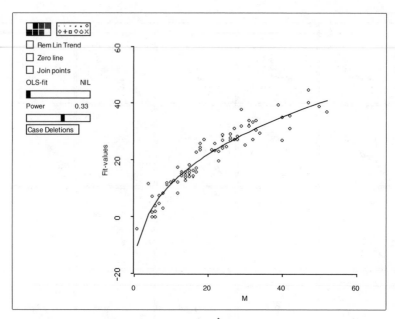

Figure 10.9. Inverse plot $\{M, \hat{M}\}$ for the mussels data.

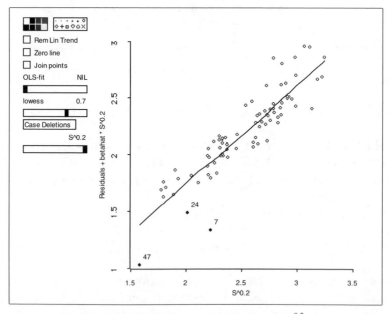

Figure 10.10. Component-plus-residual plot for $S^{0.2}$.

plot for $S^{0.2}$ based on model (10.6) is shown in Figure 10.10. It shows a strong linear trend and three relatively remote points, outliers that may have unduly influenced our conclusions. Inspect the $C+R$ plots for the other predictors. What do you conclude?

10.5 COMPLEMENTS

10.5.1 Profile Log-likelihoods and Confidence Curves

The confidences curves derived in Section 10.3 are a rescaling of a plot of the *profile log-likelihood*, which is defined by

$$L(\lambda) = -\frac{n}{2} \log(\text{RSS}_z(\lambda))$$

The value of λ that *maximizes* $L(\lambda)$ is the same as the value that *minimizes* $\text{RSS}_z(\lambda)$. The usual plot of the profile log-likelihood is of $\{\lambda, L(\lambda)\}$, as shown in the upper left of Figure 10.11. The value $\hat{\lambda}$ maximizes this curve. The second frame of Figure 10.11 is $\{\lambda, 2(L(\hat{\lambda}) - L(\lambda))\}$, obtained from the first frame by flipping the curve and changing the values on the vertical axis. In the third frame, the values on the vertical axis are replaced by their square roots. This results in a sharp point at $\hat{\lambda}$. The final frame interchanges the axes, giving the confidence curves.

10.5.2 Dynamic Probability Plot

As a complement to the Box–Cox method, one could choose λ to make a normal probability plot of residuals as straight as possible; readers unfamiliar with probability plots may want to read Section 13.4. A set of probability plots can be viewed using a slide bar, each time plotting the residuals for the regression of $z(\lambda)$ on x, as λ ranges between -2 to 2 for the power family. One would seek a λ that gives a straight plot without outliers. In addition, by marking points in the plot and following them as λ is varied, we can assess the effects of individual observations on the choice of a transformation. This plot is also produced by the "Choose Response Transform" item in the regression menu. The type of residual to use is selected in the dialog; the default is to use Studentized residuals. The different types of residuals are discussed in Section 13.5.1.

10.5.3 Transformation Families

Other families of transformations have been suggested in place of the power family, particularly for cases in which the response either is not strictly

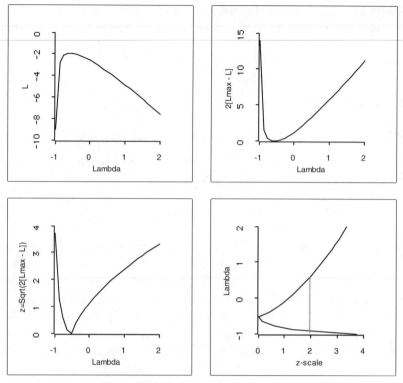

Figure 10.11. Derivation of confidence curve.

positive or is bounded on the interval 0 to 1. In the *R-code*, we include two additional families. The *modulus* family can be used when the response is not strictly positive. It was defined by John and Draper (1980) to be

$$y(\lambda) = \begin{cases} \text{sign}(y)((|y| + 1)^{\lambda} - 1)/\lambda & \lambda \neq 0 \\ \text{sign}(y)\log(|y| + 1) & \lambda = 0 \end{cases}$$

The *folded power* family (Mosteller and Tukey, 1977) is defined for data bounded on the interval from 0 to 1 to be

$$y(\lambda) = \begin{cases} (y^{\lambda} - (1 - y)^{\lambda})/\lambda & \lambda \neq 0 \\ \log(y/(1 - y)) & \lambda = 0 \end{cases}$$

Either of these families can be selected in the "Choose Response Transform" dialog.

10.5.4 References

The inverse fitted-value plot discussed in Section 10.2 for visualizing a response transformation was proposed by Cook and Weisberg (submitted for publication). Box and Cox (1964) first proposed the use of families to find a transformation of the response. They also give the wool data and various technical details concerning the derivation of the modified family; see also Cook and Weisberg (1982) and Atkinson (1985). Hernandez and Johnson (1980) showed that the Box–Cox procedure chooses transformations to make the residuals as normal as possible. The dynamic probability plot was suggested by Cook and Weisberg (1989). Confidence curves were proposed by Cook and Weisberg (1990b). They are generally useful for displaying the uncertainty in the estimates of parameters in nonlinear models where the issue of approximate normality can be more important.

The mussels data were furnished by Mike Camden, Wellington Polytechnic, Wellington, New Zealand.

EXERCISES

10.1. Beginning with the variables as transformed for model (10.6), investigate the structural dimension of the mussels regression problem. Does the structural dimension appear to be one as suggested by the analysis in Section 10.4 or is there evidence that the dimension is 2 or more?

10.2. Repeat the analysis of the mussels data in Section 10.4 after deleting cases 7, 24, and 47 from the data. These are the cases that stand out in the $C+R$ plot given in Figure 10.10.

10.3. In the wool data, we did not mention dimensionality prior to considering response transformations. Does 1D structure seem appropriate for these data? Justify your response. Continue the analysis of the wool data by replacing the response with its logarithm. Construct the inverse fitted-value plot based on the regression with $\log(y)$ as the response. Does the log transformation appear to be effective?

10.4. Investigate transforming the response in the BigMac data after replacing the four predictors *Bread*, *BusFare*, *TeachSal*, and *TeachTax* by their logarithms. Changing to logarithms achieves at least pairwise linear predictors.

10.4.1. Set up the regression using the four predictors in log scale and the response, *BigMac*. Draw the inverse fitted-value plot of {*BigMac*, $\hat{y}$}.

Does this plot suggest that transformation of *BigMac* may be necessary? Estimate the power transformation. Check on the adequacy of your estimate by using the method indicated in Exercise 10.3.

10.4.2. Construct confidence curves for the best power transformation parameter and read an approximate 95% confidence interval from the graph. Do these results agree generally with the graphical results?

10.4.3. The inverse fitted-value plot {*BigMac*, $\hat{y}$} contains four cities with the largest values of *BigMac* that are likely to be very important in determining the transformation. Identify these cities. The "Case Deletions" plot control can be used to see how deleting these cities will change the information concerning a transformation. Does deleting these cities change your view of the need to transform *BigMac*?

CHAPTER 11

Checking Models

In the last few chapters we studied methods for understanding regression problems and building models. In this chapter, we assume that we have built a model, and our goal is to decide if the model is recognizably deficient. The methods described in this chapter all depend on the residuals, which are the differences between the observed values of the response and the fitted values obtained from the model.

11.1   THE TARGET MODEL

Throughout this chapter we assume the *target model*

$$y|x = \beta_0 + \beta^T x + \varepsilon \qquad (11.1)$$

This model may have been obtained using the methods described in earlier chapters or in any other way. On one extreme, we may have decided to fit the target model without any prior data analysis because it is easy and we have no contradictory information. Alternatively, we may have used the methods described in this book to arrive at a model. In either case, our goal is to check for deficiencies.

The target model has a linear regression function and a constant variance function. The errors ε are independent of the predictors x. The target model has 1D structure because the distribution of y depends on x only through the single linear combination $\beta^T x$. All of these conditions should be checked.

11.2   THE RESIDUALS

When we fit the target model (11.1) via ols, we obtain parameter estimates $\hat{\beta}_0$ and $\hat{\beta}$ and a set of fitted values, $\hat{y} = \hat{\beta}_0 + \hat{\beta}^T x$. From the fitted values and the responses we can compute the residuals,

$$e = y - \hat{y}$$

Under the target model, the ols estimates $\hat{\beta}_0$ and $\hat{\beta}$ are *consistent* estimates of β_0 and β, meaning that we expect the differences between the estimates and the parameters to decrease as the sample size increases. How does this affect the residuals? In large samples,

$$e = y - \hat{y} = \beta_0 + \beta^T x + \varepsilon - \hat{\beta}_0 - \hat{\beta}^T x \approx \varepsilon$$

and the residuals approximate the unobservable errors. Let $e|x$ denote the residuals given that the predictors are held fixed at selected values. Under the target model the distribution of $e|x$ is essentially independent of x because the distribution of ε does not depend on x. How will this be reflected in a plot with residuals on the vertical axis and a linear combination $a^T x$ of the predictors on the horizontal axis? Regardless of the specific choice for $a^T x$, no systematic nonrandom patterns should be seen in a residual plot if the target model holds.

Analysis of residuals in small samples has two complicating factors. First, the distribution of the residuals is not completely independent of x, even under the target model, because the substitution of estimates for parameters introduces some dependence. Second, a few points that are separated from the bulk of the data can essentially determine estimates, residuals, and inferences. This problem of *influential data* is discussed in Chapter 13.

Now suppose the target model is deficient. For example, the regression function may be nonlinear when we assumed it to be linear or the variance function may not be constant when we assumed it to be so. Can we always find a residual plot that will show a clear pattern? If a pattern is found, can it be used to suggest a way to improve the model? Useful answers to both of these questions depend on the distribution of the predictors. We can say much more with linear predictors than we can for general predictors.

11.3   SELECTING RESIDUAL PLOTS

Under the target model, a 2D residual plot $\{a^T x, e\}$ should have no systematic trends. If it does have trends, then we have evidence that some aspect of the target model is questionable. If the plot $\{a^Tx, e\}$ has no systematic trends, then we have some support for the current model. The support grows as this finding is repeated for many different linear combinations a^Tx. Checking for model deficiencies when $p = 1$ or $p = 2$ can be done by inspecting the 2D or 3D plot with residuals on the vertical axis and the predictors on the other axes. Failing to find a systematic relationship between the residuals and the predictors supports the target model.

Model checking is harder when $p > 2$ because we cannot inspect the full $(p+1)$-dimensional plot of residuals versus predictors. We can inspect 2D and 3D residual plots of the form $\{a^Tx, e\}$ and $\{a^Tx, e, c^Tx\}$ where a^Tx and c^Tx are linear combinations of the predictors. If we find a systematic pattern in any of these plots, we have evidence of model deficiency. Inspecting residual plots for all possible choices of a and c cannot be done in practice. The hard part is knowing which plots to choose and when to stop. We now describe a rationale for choosing a few residual plots that can provide good support for the target model when no systematic patterns are present.

11.3.1   The Setup

Think of the process that led to the target model. Through the modeling procedure, we obtain the following classification of predictors:

- A set w of linear predictors. In the Big Mac data, for example, approximately linear predictors are obtained by taking logs of the four predictors. In the mussels data, only one of the original predictors needed to be transformed. Getting exactly linear predictors is not very important, but removing strong nonlinear trends is important.

- A set $x = (x_1, \ldots, x_p)^T$ of p predictors produced by modeling. The key assumption is that each of the predictors in x can be written as a function of one or more of the linear predictors in w. These could include derived variables such as quadratics and cross products that are functions of other predictors. Predictors could also be empirically defined transformations based on smoothers, as was done in Section 9.3.

The target model (11.1) uses the variables in x as predictors, so the model need not be restricted to linear predictors.

11.3.2   Linear Predictors

Let's consider first the case $x = w$, so all the predictors in the model are linear predictors. Suppose in addition the assumption of 1D structure in the target model holds. The 1D estimation result of Section 7.4.1 tells us that the plot $\{\hat{y}, y\}$ provides a summary, without loss of essential information. The summary plot can be used to visualize the regression function and the variance function from the general 1D model given by (7.3). Under the target model, the regression function is linear and the variance function is constant.

Removing the linear trend from $\{\hat{y}, y\}$ gives the residual plot $\{\hat{y}, e\}$. Under the target model this plot will have no systematic features. If it is curved, then the regression function $E(y|x)$ is nonlinear. If the plot is fan-shaped or otherwise shows systematic changes in variation across the plot, then $var(y|x)$ is not constant.

Curvature in the plot $\{\hat{y}, e\}$ may indicate the need to transform either the response or the predictors. Response transformations are consistent with 1D structure, but predictor transformations are not. A diagnostic graph for transforming the response is just the plot $\{y, \hat{y}\}$, as described in Section 10.2. To study predictor transformations, use the $C+R$ plots discussed in Section 9.1.

In summary, with linear predictors and 1D structure, the single 2D residual plot $\{\hat{y}, e\}$ is usually enough to check on the adequacy of the target model (11.1) against the general 1D model (7.3). If the model is found to be lacking, other plots may be needed to decide what to do next. If the structural dimension of the regression problem is greater than 1, this plot will not be adequate because the summary plot $\{y, \hat{y}\}$ can miss essential information. Plots that can be used to detect incorrect structural dimension are discussed in Section 11.4.

As an example, the plot $\{\hat{y}, e\}$ for the Big Mac data with *BigMac* as the response and the four predictors in log scale is shown in Figure 11.1. This plot has three primary features. First, Mexico City is poorly fit. Second, the plot appears curved. This is confirmed by the *lowess* smooth, which was drawn after using the "Remove Selection" item to remove the point for Mexico City. The item "Show All" was then used to restore Mexico City to the plot. Finally, variability seems to increase as we move to the right. These features indicate a problem with the target model. The most likely remedy is to transform the response; see Exercise 10.4.

As another example, return to the New Zealand birth data described in Exercise 4.3. Choose *Term*, the length of pregnancy in weeks, as the response. The other three variables, *BirthWt*, *Age* of mother, and *Sex*, are the predictors. The summary residual plot $\{\hat{y}, e\}$ is shown in Figure 11.2. The striking feature in this plot is a set of parallel stripes with negative slopes. Are the stripes a cause for concern?

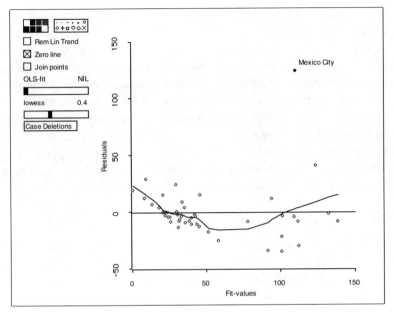

Figure 11.1. Residuals versus fitted values for the Big Mac data, with the response untransformed.

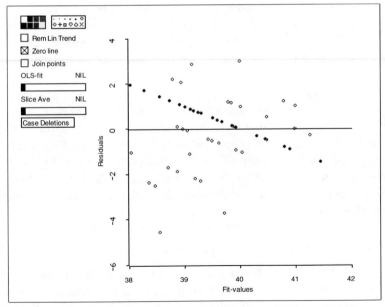

Figure 11.2. Residual plot with *Term* as the response for the Wellington birth data.

In general, striped patterns in residual plots are caused by discreteness in the response. *Term* is an integer between 36 and 43 weeks; for example, the selected points in the plot all have *Term* equal to 40 weeks. Since all the cases with *Term* equal to 40 weeks have different values of the predictors, they have different fitted values, and so the residuals vary along a line. We have previously seen a similar phenomenon in Figure 2.4, where many values of the response close to zero were observed, leading to a sharp edge at the lower left of that plot.

To evaluate a plot for lack of fit of a model, stripes are generally not important, since they are due to discreteness, not to lack of fit. There is little evidence in the residual plot against the target model.

11.3.3 General Predictors

With linear predictors and 1D structure, the single residual plot $\{\hat{y}, e\}$ is usually all that is needed to check the adequacy of the target model (11.1) against the general 1D model (7.3). Without linear predictors, this plot may miss dependence of the residuals on x because the 1D estimation result of Section 7.4.1 does not apply.

This can be demonstrated by using one of the items from the `demo-3d.lsp` demonstration in the `R-data` folder. Load this file, and select the item "Nonlinear Predictors, Nonlinear Model" from the menu. Set the predictors to be x_1 and x_2 and set the response to be y. This example has a nonlinear response function and nonlinear predictors, as discussed in Section 7.4.2. Draw the plot $\{\hat{y}, e\}$, as shown in Figure 11.3. The plot gives hints that something may be wrong. The bulk of the data on the left exhibits a negative linear trend. This may be a distortion caused by influential points on the right of the plot. Nothing in the plot suggests that a better model is quadratic rather than linear in the predictors.

We have only two predictors in this example so a 3D plot with residuals on the vertical axis is the only residual plot that is needed to check the model. One particular view of this 3D plot is the same as Figure 11.3. While rotating the 3D plot, a strong pattern is visible. The 2D residual plot $\{\hat{y}, e\}$ missed key information about the dependence of e on x because of the nonlinear dependence between the predictors.

11.3.4 Restricting to Linear Predictors

Suppose w is the set of linear predictors in a regression problem and x consists of transformations of w. If the distribution of $e|w$ does not depend on the value of w, then the distribution of $e|x$ cannot depend on the value of x. If the distribution of $e|w$ does depend on the value of w, then there

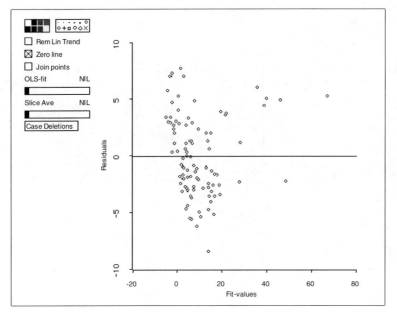

Figure 11.3. Plot of ols residuals versus fitted values for the "Nonlinear Predictors/Nonlinear Model" demonstration.

is a model deficiency by definition. Thus, we only examine residual plots with linear combinations of the linear predictors w on the horizontal axes. This can reduce the number of residual plots we need to check.

Return to the ethanol data discussed in Sections 7.3 and 9.3. We started with two predictors C and E which we found to be approximately linear predictors, so $w = (C, E)^T$. In Section 7.3 we decided that $NOx|w$ exhibited 2D structure. In Section 9.3 we used $C+R$ plots to choose predictor transformations. This led to the target model (9.8),

$$NOx = \beta_0 + \beta_1 C + \beta_{11} C^2 + \beta_2 TE + \varepsilon$$

where TE is the transformation of E chosen via the *lowess* smooth in Figure 9.6. The model has three predictors $x = (C, C^2, TE)^T$ and is linear in x. Predictor transformations may have reduced the structural dimension to 1, but since x is not a set of linear predictors and has more than two elements, we do not have a direct check on dimension.

To study lack of fit, we need only a single 3D residual plot $\{E, e, C\}$ of the linear predictors. To draw this plot, you must first reconstruct the predictors in x: Fit the regression of NOx on C and E, and use the $C+R$ plot for E and the *lowess* smoother to extract TE. Then use the "New Model. . ." item to fit the model with NOx as the response and C, C^2, and

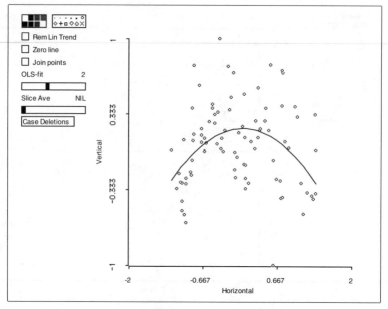

Figure 11.4. Ethanol data residual plot.

TE as predictors, creating C^2 by using the "Transform..." button. The plot can now be drawn by using the "Plot of..." item. After rotating the plot, what do you conclude about the adequacy of the model? Strong nonlinear trends are visible in the residuals, so systematic variation in the response remains unexplained by the predictors. In particular the distribution of $e|w$ depends on the value of C. One extracted 2D view from the 3D plot is shown in Figure 11.4 with a quadratic trend superimposed. The model does not fully explain all the systematic variation in *NOx*. The most likely problem is with the assumption of 1D structure.

Rotate $\{E, e, C\}$. Since C consists of only five values, the plot consists of five separate plots, one for each value of C. A helpful way to view these five curves is to draw the plot $\{E, e\}$ and use the "Slicer..." control from the plot's menu to slice the plot on C. All five plots are reasonably linear with slope changing from negative for small values of C to positive for large values of C. This behavior is characteristic of an interaction between C and E and suggests that adding a term of the form $C \times TE$ may improve the current model.

The $C+R$ plots took us a long way toward a useful analysis of the ethanol data, but systematic features remain because the procedure does not allow for transformations that depend on more than one predictor. The elaboration of the model suggested here is relatively minor and will result in only relatively small changes in predictions. The simpler model without an interaction may be completely adequate for many purposes.

11.4 *ALP* RESIDUAL PLOTS

With more than two linear predictors in w, the residuals cannot be plotted against all elements of w at once. We are again faced with the problem of deciding which residual plots to view. Fortunately, there is a useful strategy based on one 2D residual plot for each linear predictor in w.

11.4.1 *ALP* Residual Plots in Two Dimensions

Let w_2 denote a single linear predictor and collect the remaining linear predictors into a vector w_1. The usual ols residuals from the regression of w_2 on w_1 will be denoted by $e(w_2|w_1)$. We still use the notation e to denote the residuals from the target model (11.1). The plot $\{e(w_2|w_1), e\}$ is used to determine if the distribution of $e|w$ depends on the single predictor w_2 in the same way that any residual plot is used to assess model adequacy. A systematic pattern is an indication of a model deficiency, while a random pattern supports the current model.

We will refer to plots of the form $\{e(w_2|w_1), e\}$ as *adjusted linear predictor residual plots*, or *ALP* residual plots for short. The *adjustment* indicates that the predictor w_2 has been adjusted for the remaining predictors w_1 by using the residuals $e(w_2|w_1)$ rather than w_2 itself. This avoids confusing the effects of w_2 with the effects of the other linear predictors. The term *linear predictors* in the name refers to using only the linear predictors w in the plots, rather than the predictors x that may be derived from them.

We have one *ALP* residual plot for each linear predictor. If the *ALP* plot for w_2 does not show a systematic pattern, then we have an indication that the distribution of $e|w$ does not depend on the value of w_2. If none of the *ALP* plots show a systematic pattern, then we have an indication that the distribution of $e|w$ does not depend on the value of w, which translates into support for the current model. Any systematic pattern in an *ALP* plot indicates a model deficiency.

We use the mussels data to illustrate 2D *ALP* plots. Our analysis in Section 10.4 led to the target model

$$M^{1/3} = \beta_0 + \beta_1 L + \beta_2 W + \beta_3 H + \beta_4 S^{0.2} + \varepsilon \qquad (11.2)$$

with $w^T = (L, W, H, S^{0.2})$ forming a set of linear predictors. In this

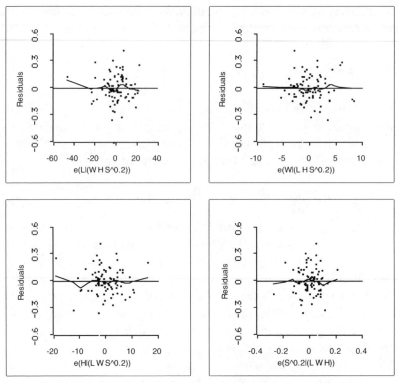

Figure 11.5. *ALP* plots for the four predictors in (11.2) for the mussels data.

example the linear predictors w and the model predictors x are the same, $w = x$. There are four 2D *ALP* plots, one for each linear predictor.

To construct the *ALP* plots, load the file `mussels.lsp` from the R-data folder, transform M and S as used in the model, and set up the regression. Select the item "ALP Res Plots–All 2D" from the "Mussels" menu. You will then be presented with a dialog to select the linear predictors. After selecting the linear predictors in $w^T = (L, W, H, S^{0.2})$, a 2D plot will appear on the screen with an extra slide bar that is used to choose the linear predictor on the horizontal axis. The four *ALP* plots for the mussels data are shown in Figure 11.5, each with a horizontal line at zero and a *lowess* smooth added. Aside from a few outlying points, none of the *ALP* plots indicates a model deficiency, lending support to (11.2).

11.4.2   *ALP* Residual Plots in Three Dimensions

Variation in the data can mask systematic features in an *ALP* plot, even if smoothers and other plot enhancements are used. A way to gain resolution is to use a 3D *ALP* plot, $\{e(w_2|w_1), e, e(w_3|w_1)\}$. Here w_2 and w_3 are individual linear predictors, and the remaining linear predictors are collected into the vector w_1. A 3D *ALP* plot is interpreted like a 2D *ALP* plot. For example, if $\{e(w_2|w_1), e, e(w_3|w_1)\}$ appears as a random point cloud with 0D structure, then we have an indication that the distribution of $e|w$ does not depend on the values of w_2 and w_3. Three-dimensional *ALP* plots are more powerful than 2D *ALP* plots because they allow detection of 2D structure, and variation around any systematic trends will be smaller. As with all 3D plots, 3D *ALP* plots should be used sparingly because inspecting a 3D plot can take considerably longer than inspecting a 2D plot. Considering all possible 3D *ALP* plots is certainly not something that would normally be done in practice. A good use of 3D *ALP* plots is to study pairs of linear predictors that have suggestive or nonrandom 2D *ALP* plots. A series of 3D *ALP* plots might be used routinely as a further diagnostic check provided each linear predictor is constrained to appear in only one 3D plot. The total number of 3D plots needed is then about half the number of linear predictors, which may be manageable with many predictors.

Continuing with the mussels data, select the item "ALP Res Plot–3D" from the regression menu and specify the linear predictors in the next dialog, just as for a 2D *ALP* plot. An initial 3D plot is drawn using the first two of the linear predictors as w_2 and w_3; the linear predictors used in the plot can be changed by pushing the "Change Predictors" button on the 3D *ALP* plot. Figure 11.6 shows one 2D view of the 3D *ALP* plot for H and

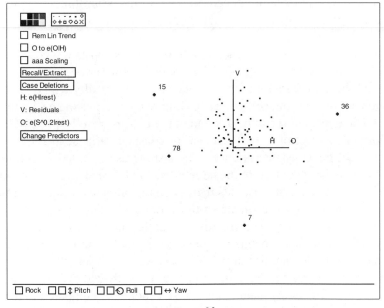

Figure 11.6. 3D *ALP* plot for H and $S^{0.2}$ in (11.2) for the mussels data.

$S^{0.2}$. Aside from a few remote points that require further study, this plot supports the target model.

11.5 NONCONSTANT VARIANCE

The target model (11.1) has a constant variance function. The residuals can be used both to test for nonconstant variance and to help decide on the form of the variance function. One general model for nonconstant variance is

$$\text{var}(y|x) = \sigma^2 \exp(\alpha^T x) \qquad (11.3)$$

where α is a vector of parameters and σ^2 is an unknown positive constant that gives the variance of the response when $x = 0$. When $\alpha = 0$, the variance function is constant. The exponential function is used to insure that the variance function is always positive, but the exact functional form is not very important. If the target model (11.1) holds but with variance function given by (11.3), we will have 2D structure because two linear combinations are required to model the dependence of y on x: $\beta^T x$ for the regression function and $\alpha^T x$ for the variance function.

An important special case of (11.3) is

$$\text{var}(y|x) = \sigma^2 \exp(\gamma \beta^T x) \qquad (11.4)$$

We now have 1D structure because both the regression function and the variance function depend on the same linear combination $\beta^T x$. Many real-world phenomena exhibit variance that changes with the mean response, as modeled by (11.4). The variance function (11.4) is constant when $\gamma = 0$.

If we knew α, we could visualize nonconstant variance in a plot of $\{\alpha^T x, e\}$, because variability would change systematically with $\alpha^T x$. Perception of nonconstant variance can be improved by replacing e with $|e|^{1/2}$. We would expect to see a linear trend in the plot $\{\alpha^T x, |e|^{1/2}\}$ if $\alpha \neq 0$.

For (11.3) a test of $\alpha = 0$ is a test for constant variance against an alternative of nonconstant variance. The test we use is called a *score test*. To compute the score test, fit the target model (11.1) under the assumption of constant variance and use this regression to compute the squared residuals e^2. Each e^2 is an estimate of the variance for that case, and so the squared residuals contain information about the variance function. Should the variance function depend on x, the regression of e^2 on x would account for much of the variability in e^2. The test statistic is just the ols regression sum of squares for the regression of e^2 on x divided by a scale factor, $2(\sum e^2/n)^2$. To get a p-value, this statistic should be compared to the χ^2

distribution with degrees of freedom equal to the number of components in α.

To test $\gamma = 0$ in model (11.4), fit the ols regression of e^2 on $\hat{y}$, the fitted values from the fit of the target model, and then compute the statistic in the same way. The resulting statistic has one degree of freedom.

11.6 **EXAMPLE:** *CASUARINA* **DATA**

Seeds from a tropical tree called *Casuarina cunninghamii* were collected from two seed sources. Six plants from each seed source were grown with fertilizer and the other six without. After four years, the diameter was measured at 65 cm above ground. The trees were then cut and weighed. The goal for this example is to study the relationship between the response weight W and the three predictors: diameter D, the use of fertilizer F, and seed source S.

Load the file `casuarin.lsp` from the `R-data` folder and set W to be the response and D, S, and F as predictors. The latter two are indicator variables, with S taking the values 0 and 1 to indicate the seed sources and F taking the value 0 to indicate no fertilizer and 1 to indicate that fertilizer was used.

How might we expect W to vary as a function of D? Elementary geometric considerations suggest that W should equal volume of wood × wood density per unit volume. Further, if a tree were a cylinder, volume = $(\pi/4)\times$ height $\times D^2$. Putting these two together, we get

$$\text{E}(W\,|\,D, \text{height}, \text{density}) \approx \pi/4 \times \text{density} \times \text{height} \times D^2 \qquad (11.5)$$

This relationship depends on height and density, which are unknown, and it does not depend on F or S. Nevertheless, it suggests that we might need to use D^2 as a predictor, not just D. We can explore the need to transform D by using the C+R plot for D from the regression of W on D, S, and F, as shown in Figure 11.7. This plot is curved and the curve is closely matched by fitting a quadratic. This suggests fitting a model with D^2 added to the original three predictors. Our initial working model is thus

$$W\,|\,x = \beta_0 + \beta_1 D + \beta_{11} D^2 + \beta_2 S + \beta_3 F + \text{error}$$

Draw the plot of $\{\hat{y}, e\}$ based on the working model. The points with the five largest absolute residuals correspond to larger fitted values and to the five heaviest trees. Increasing variance as a function of D is also apparent in Figure 11.7. This might lead one to suspect that larger trees are more

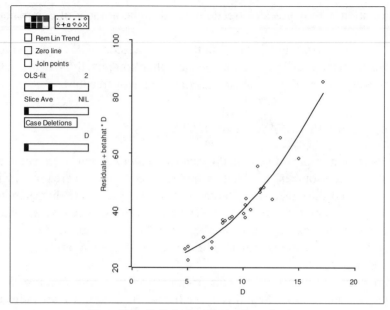

Figure 11.7. $C+R$ plot for D in the *Casuarina* data.

variable. To explore this further, select the item "Nonconstant Variance Plot" from the regression menu for the working model. This will produce the plot shown in Figure 11.8. The vertical axis in this plot is $|e|^{1/2}$. The plot is initially based on model (11.4) with the variance a function of the mean, and so the initial horizontal axis is just $\hat{y}$. The general trend in the plot is increasing to the right, suggesting that variance increases with the mean. This is confirmed by the score test, given at the top of the plot. The value of the statistic is 9.23, with one degree of freedom, giving a p-value of 0.002.

Individual points separated from the main trend may determine the value of the score statistic. These can be identified in the graph and deleted with the "Case Deletions" item. The plot will then be automatically updated. In Figure 11.8, 5 of the 24 points are generally to the right and above the rest of the points. Deletion of all five of these cases makes the evidence for nonconstant variance disappear. In such a small data set deleting the five largest trees is undesirable because it limits inference to smaller trees. Restore all the data using the "Restore All" item from the "Case Deletions" pop-up menu to continue analysis.

An additional plot control called "Change Model" appears on the nonconstant variance plot. This control is used to change the predictors

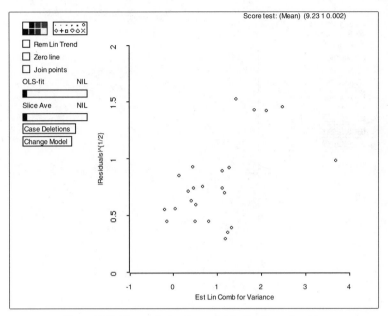

Figure 11.8. Nonconstant variance plot for the *Casuarina* data.

in the variance function. If you push the mouse button in this control, a
dialog appears that allows choice between modeling variance as a function
of the mean or of one or more of the predictors. To specify a variance
model that is a function of D only, double click on D, and then push "OK."
To return to model (11.4), push the button for the mean. To get the most
general model, select all four predictors. The result of this last selection
is shown in Figure 11.9. The horizontal axis in this plot is an estimate of
$\alpha^T x$.

The score test shown in Figure 11.9 is equal to 10.24, compared to
9.23 for variance as a function of the mean. Since modeling variance as a
function of the mean is a submodel of modeling variance as a function of
all the predictors, we can subtract the two score tests, $10.24 - 9.23 = 1.01$
with $4 - 1 = 3$ degrees of freedom, to get an approximate χ^2 test for
comparing the two models. The p-value for this test is computed by typing

```
> (- 1 (chisq-cdf 1.01 3))
0.798832
```

This subtracts the area under the $\chi^2(3)$ density to the left of 1.01 from 1,
giving the p-value close to 0.8. The large p-value suggests that variance

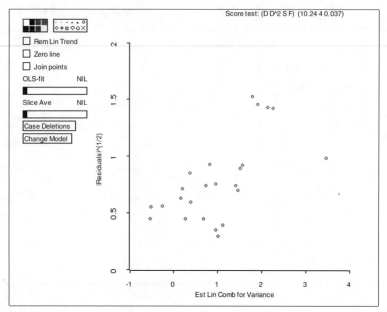

Figure 11.9. Nonconstant variance plot as a function of the predictors for the *Casuarina* data.

can be modeled as a function of the mean only and that 1D structure is adequate.

An informative exercise is to associate the points in Figure 11.8 with the levels of S and F. An easy way to do this is to draw a plot of $\{S, F\}$. This plot consists of only four points (why?); as you select the points in this graph, the corresponding points in the nonconstant variance plot will be highlighted. What do you learn from this?

We have discovered that the variability in weight appears to be an increasing function of the mean. What shall we do next? Let's fix S and F and consider only D. One possible remedy for nonconstant variance is transforming the response and possibly the predictors. Returning to equation (11.5) and taking logarithms, we get

$$E(\log(W)|D, \text{height}, \text{density}) \approx \log(\pi/4) + \log(\text{density})$$
$$+ 2\log(D) + \log(\text{height})$$

The right side of this equation includes three predictors, density, height, and diameter D, but of these only D is recorded in the data. To make any progress, we need to average over the predictors we have not observed to get a regression function for D only. We did a similar calculation in equation (6.4), but for a different purpose. We find

$$E(\log(W)|D) \approx \log(\pi/4) \;+\; E(\log(\text{density})|D)$$
$$+\; 2\log(D)$$
$$+\; E(\log(\text{height})|D)$$

This equation has only one predictor on the right, namely $\log(D)$, but it also depends on the regression functions for the logarithms of density and height given D. We have no data to estimate these regression functions, so we must make an informed guess about them. Suppose that wood density is independent of the diameter of the tree, so $E(\log(\text{density})|D)$ is constant for all trees in a seed source–fertilizer combination. Call this constant d. This leaves height. We can hypothesize about the regression function for height on diameter. Suppose

$$E(\text{height}|D) = m_0 D^{m_1}$$

If $m_1 = 1$, then the regression function is linear. If $m_1 = 0$, then the regression function is constant. Other values of m_1 provide nonlinear relationships between height and diameter. We can write $E(\log(\text{height})|D) \approx \log(m_0) + m_1 \log(D)$. Combining all these, we get a regression function for $\log(W)$:

$$E(\log(W)|D) \approx \log(\pi/4) + d + \log(m_0) + (2 + m_1)\log(D) \quad (11.6)$$

The intercept is a function of d, m_0, and constants, so neither d nor m_0 is estimable. The slope is a function of m_1 alone, so m_1 is estimable.

The model so far is for fixed values of S and F. These can be included in the model in many ways. A simple model would include S and F linearly. This is equivalent to assuming that m_1 is the same for all combinations of S and F. We could allow m_1 to vary with S and F by including the interaction terms $S \times \log(D)$ and $F \times \log(D)$ in the model.

This suggests fitting a model with $\log(W)$ as the response and with predictors S, F, and $\log(D)$. When this model is fit, the ols coefficient estimate for $\log(D)$ is 2.155 with a standard error of 0.095, and the t statistic for testing $m_1 = 0$ is $(2.155 - 2)/0.095 = 1.63$. Comparing this to the $t(20)$ distribution gives a two-tailed p-value of about 0.11, so these results are reasonably consistent with $m_1 = 0$. At least for these four-year-old trees, $E(\text{height}|D)$ appears to be independent of diameter D. Has the change to the log scale corrected the problem with nonconstant variance? Shown in Figure 11.10 is the nonconstant variance plot as a function of seed source S. Modest evidence of nonconstant variance remains, with trees from seed source 0 more variable than those from seed source 1.

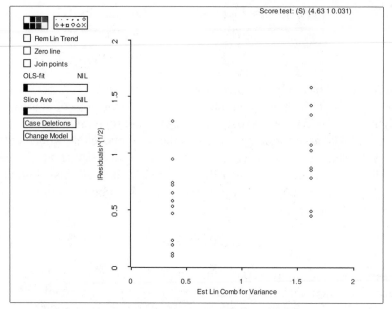

Figure 11.10. Nonconstant variance plot for seed source for the *Casuarina* data.

Inspect the *ALP* residual plots, specifying all three predictors as linear predictors. Does the *ALP* plot for *S* support the finding of nonconstant variance as a function of *S*?

Although nonconstant variance has not completely disappeared, modeling variance as a function of seed source only is considerably simpler than modeling it as a function of several predictors or of the mean. Indeed, the data suggest that trees from seed source 1 are both heavier and more variable than are trees from seed source 0.

11.7   COMPLEMENTS

11.7.1   Residuals

A few of the residuals may not behave like the corresponding errors ε, even when sample size is large and the target model holds. Consider simple linear regression with $n - 1$ observations taken at $x = 0$ and only one observation taken at $x = 1$. No matter how large the value of n, the residual for the observation at $x = 1$ will always be equal to zero, regardless of the value of $\varepsilon|(x = 1)$. The condition needed to guarantee that e behaves like the corresponding error is that as the sample size increases, the maximum value of the leverages, which we will define in Section 13.5.1, must approach zero. This condition is usually, but not always, satisfied.

Huber (1981, Section 7.2) provides a technical discussion. Searle (1988) provides additional discussion of stripes in residual plots.

11.7.2   Regression Functions and Linear Predictors

Even with linear predictors and 1D structure, the residual plot $\{\hat{y}, e\}$ may miss relevant information. This will occur, for example, if the regression function is a symmetric quadratic because then the constant c in the 1D estimation result of Section 7.4.1 is equal to 0. A discussion of this in the context of *SIR* is given in Cook and Weisberg (1991a). Further discussion of this issue, and much of the theory used in this chapter, is given by Cook (1994) and Cook and Wetzel (1993).

11.7.3   Nonconstant Variance

The score test for nonconstant variance was given by Breusch and Pagan (1979) and by Cook and Weisberg (1983). The results of Chen (1983) suggest that the form of the variance function is not very important. Hinkley (1985) states that the difference of score tests can be compared to a χ^2 to compare nested models. The use of $|e|^{1/2}$ in the nonconstant variance plots is based on results given by Hinkley (1975). Other relevant references on variance modeling and diagnostics for variance problems include Carroll and Ruppert (1988) and Verbyla (1993).

The *Casuarina* data were provided by Ross Cunningham. The lettuce data in Exercise 11.5 are taken from Cochran and Cox (1957, p. 348).

EXERCISES

11.1.   *11.1.1.*   Conduct an analysis of the residuals from the model proposed for the Australian athletes data in Section 8.2. The response is *LBM* and the predictors are *Sex*, *Wt*, *Ht*, *RCC*, *Sex* × *Wt*, *Sex* × *Ht*, and *Sex* × *RCC*. Include an analysis of nonconstant variance as described in Section 11.5, and use 2D and 3D *ALP* plots.

11.1.2.   Apply *SIR* to the regression of *e* on *Sex*, *Wt*, *Ht* and *RCC*, where *e* denotes the ols residuals from the model of the last section of this exercise. Suppose the BodyMass is the name of the regression you fit in the last part of this exercise. First, enter the command

```
(send BodyMass :add-data (send BodyMass :residuals) "e")
```

This will add the residuals to the data set. Use the "New Model..." item from the regression menu to set up the model you need, and then use the "Inverse Regression" item from the new model's menu to compute *SIR*. Compare the results you get from *SIR* to the results you obtained in the last section of this exercise.

11.1.3. The predictor *Sex* is an indicator variable. Does this cause problems for the assumption of linear predictors needed for *SIR*, both in general and in specific for this exercise?

11.2. Conduct a residual analysis of the model you obtained for the haystack data in Exercise 7.8. Improve the model if necessary.

11.3. Conduct a residual analysis of the model you obtained for the land rent data in Exercise 6.7. Improve the model if necessary.

11.4. Complete the analysis of the *Casuarina* data. What do the *ALP* plots suggested at the end of Section 11.6 show? Fit a model you might use that allows for the relationship between density and height to depend on seed source *S*. What is the effect of *F*?

11.5. The data in file `lettuce.lsp` in the `R-data` folder are the results of a central composite design on the effects of the log concentration of the minor elements copper *Cu*, molybdenum *Mo*, and iron *Fe* on the growth of lettuce in water culture. The response is lettuce yield, *y*. The sample size is only 20, so seeing trends in graphs may be a bit difficult.

Set up the regression with *y* as the response and the other three variables as the predictors. Examine the scatterplot matrix of the four variables. From the scatterplot matrix, describe the design. Use the methods described in this book to model *y* as a function of the predictors, and then use the appropriate graphical method to assess the lack of fit. Summarize your results. Hint: Central composite designs are used to find the maximum or the minimum value of a response surface, so it is likely that $E(y|Cu, Mo, Fe)$ is not a monotonic function of the predictors.

11.6. Try adding the $TE \times C$ interaction to the model for the ethanol data, as discussed at the end of Section 11.3.4. Does this completely remove all lack of fit of the model? If not, what problems remain? What might you try to do next to improve the model?

CHAPTER 12

Assessing Predictors

In this chapter, we present two graphical methods for visualizing the contribution of each predictor in turn to the fit of an adequate target model. The graphs are called added-variable plots and *ARES* plots. Added-variable plots require linear predictors, but ARES plots do not.

12.1 ADDED-VARIABLE PLOTS

We begin again with an adequate target model,

$$y|x = \beta_0 + \beta_1^T x_1 + \beta_2 x_2 + \varepsilon \qquad (12.1)$$

with a constant variance function. The predictors x have been partitioned into two parts, a vector x_1 of $p-1$ predictors and a single predictor x_2. We seek a plot that shows the contribution of x_2 to this model. We could try the partial response plot $\{x_2, y\}$, but we know from Section 6.4.1 that a partial response plot may show no systematic patterns when x_2 does contribute significantly to the regression, or it may show a relationship when x_2 is not needed once x_1 is included. The problem is that the partial response plot does not adjust for the contributions of x_1. A plot that does the appropriate adjustment is the added-variable plot.

Let $e(y|x_1)$ denote the residuals from the ols regression of y on x_1 and let $e(x_2|x_1)$ denote the residuals from the ols regression of x_2 on x_1. An added-variable plot is defined as $\{e(x_2|x_1), e(y|x_1)\}$. Beginning with the

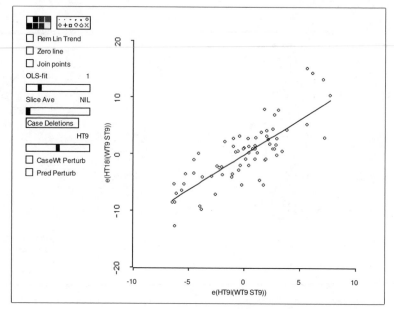

Figure 12.1. Added-variable plot for *HT9* from the Berkeley Guidance Study for girls.

partial response plot $\{x_2, y\}$, an added-variable plot is obtained by replacing each variable by its residuals from the ols regression on x_1. This process removes the contribution of x_1 to the fit of the target model.

As an example we return to the Berkeley Guidance Study for girls as described in Section 7.5. The response is *HT18*, the height of a girl at age 18, and the predictors are the height, weight, and strength at age 9, *HT9*, *WT9*, and *ST9*, respectively. We previously concluded that the target model (12.1) with $x = (HT9, WT9, ST9)^T$ is reasonable for these data. In the *R-code*, select the item "AVP–All 2D" from the regression menu. The 2D plot that results is the added-variable plot for one of the predictors in the model. An additional slide bar and two additional buttons appear on the plot. Clicking in the slide bar changes the predictor in x that is to be the added variable. Discussion of the buttons is deferred to Section 13.5.2. The added-variable plot for *HT9* is shown in Figure 12.1 along with the ols line for the plot.

Use the slide bar to cycle through the three added-variable plots for the example. Each plot enables us to visualize the contribution of the corresponding predictor to the regression after the other two predictors. The plots can be interpreted like response plots in simple linear regression. In particular, if the target model holds, then *each added-variable plot must exhibit linearity*. Nonlinearities mean that the target model is deficient in some way. Nonlinearities also imply that it is premature to use added-variable plots to visualize the contributions of predictors since this technique is restricted to situations where the target model is adequate and the predictors are linear. There are no obvious nonlinear trends in any of the added-variable plots for the Berkeley Guidance Study, supporting the idea that the target model may be reasonable for these data.

The strength of a linear trend in an added-variable plot for x_2 is an indication of the strength of the contribution of x_2 after the other predictors in x_1. The added-variable plot for *HT9* in Figure 12.1 shows a fairly strong linear trend, indicating that this variable contributes substantial information to the regression beyond that contributed by the other two variables.

Lack of a systematic relationship in an added-variable plot means that x_2 provides no additional explanatory power after the other variables, but it does not mean that y and x_2 are independent. Indeed, the partial response plot $\{x_2, y\}$ can show a strong linear trend even if x_2 provides no additional explanatory power after x_1. When this happens, x_1 will be a good predictor of x_2.

12.1.1   Properties of Added-Variable Plots

Added-variable plots are closely related to some regression calculations. Consider fitting a line by ols to the points in an added-variable plot, as shown in Figure 12.1. As long as the intercept is in the target model, the estimated intercept in the plot will be zero. The estimated slope in the added-variable plot will always be the same as the estimated coefficient $\hat{\beta}_2$ in the ols fit of the target model. The t-statistic for the hypothesis $\beta_2 = 0$ in the original regression is a constant times the t-statistic for testing the slope to be zero in the added-variable plot regression. Because of this relationship, an added-variable plot provides a visualization of the corresponding t-test.

Finally, the residuals in an added-variable plot are identical to the residuals e from the fit of the target model, so we can write

$$e = e(y|x_1) - \hat{\beta}_2 \times e(x_2|x_1) \tag{12.2}$$

These results are pursued in Exercise 12.1.

12.1.2   Some Extreme Added-Variable Plots

In this section we discuss three extreme cases of added-variable plots. While these cases may rarely occur in practice, they provide additional

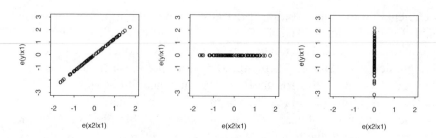

Figure 12.2. Three extreme added-variable plots.

understanding of added-variable plots. The three cases are depicted in Figure 12.2.

In the left frame of this plot, all the points lie exactly on a straight line with nonzero slope. What does this tell us about the fit of the target model (12.1)? Since the points fall exactly on a line, the residuals from the added-variable plot regression are all zero. Using this fact with equation (12.2) tells us that the residuals from the target model (12.1) must all be zero and all the data fall exactly on a plane in p dimensions. Adding x_2 to the model gives a perfect fit, so x_2 is useful even after the contributions of the variables in x_1. In less extreme cases, x_2 is important if its added-variable plot is strongly linear, with small residuals.

Suppose next that the points in an added-variable plot lie exactly on a straight line with zero slope, as in the second frame of the plot. In this case we must have $e(y|x_1) = 0$ for every case in the data. There is no reason to consider adding x_2 because the regression of y on x_1 explains all the variation in the response. In less extreme cases, only the slope in the plot needs to be zero for the predictor to be unimportant.

The third frame illustrates the extreme case with $e(x_2|x_1) = 0$ for all cases in the data. This means that x_2 is an exact linear combination of the other predictors. We need not include x_2 in the model because any information that x_2 may contain about y is already accounted for by x_1. The *R-code* would print the word *aliased* in place of an estimated coefficient for x_2. This situation is often identified by the term *colinearity*. Exact colinearity is rarely observed because of the finite arithmetic used on computers. Approximate colinearity, where one of the predictors is almost a linear combination of the others, is quite common in some areas of application. It can be diagnosed if the range of the values of $e(x_2|x_1)$ is tiny relative to the range of the values of x_2.

12.1.3   Three-Dimensional Added-Variable Plots

An added-variable plot can be generalized to three dimensions. Let x_2 and x_3 be individual predictors, collecting the remaining predictors into x_1. The *3D added-variable plot* is then$\{e(x_2|x_1), e(y|x_1), e(x_3|x_1)\}$, where $e(x_3|x_1)$ are the residuals in the ols regression of x_3 on x_1. As with the *ALP* plot, the 3D added-variable plot gains resolution over its 2D cousin by reducing background variation. They can be interpreted in much the same way as 3D response plots. Nonlinearities, fan shapes, and evidence of 2D structure all indicate that the target model may be inappropriate.

12.2   *ARES* PLOTS

Like the added-variable plot, the *ARES* plot provides a method of visualizing the importance of a predictor in the fit of a target model. The new ingredient is the method of visualization: The *ARES* plot uses animation to convey information. Suppose the importance of x_2 in the target model is at issue, where $x^T = (x_1, x_2)$. We have two models to compare, the target model (12.1) with predictors x and a subset model with predictors x_1. Imagine a plot that provides a smooth transition from the subset model to the target model. By varying the "fraction" of x_2 added while viewing an appropriate plot, we may get useful insights into the difference between the two models. The name of the plot is *ARES*, which is an acronym for *A*dding *RE*gressors *S*moothly.

12.2.1   *ARES* Plots for a Single Predictor

Return to the Berkeley Guidance Study for girls, and set up the regression with *WT18* as the response and all the age 9 and the remaining age 18 variables as predictors. Select the item "ARES Plot" from the regression menu. In the resulting dialog, double click on *LG18* to select this predictor and then push "OK." The graphical display will provide a visualization of the effects of adding *LG18* to a model that contains all remaining predictors. In the generic notation, $x_2 = LG18$, and x_1 is the vector of all remaining predictors.

The initial view of this plot is {fitted values, residuals} computed from the ols fit of the subset model with predictors x_1, as shown in Figure 12.3. The usual *R-code* plot controls appear on this plot, with an additional slide bar labelled "lambda." Informally, λ can be thought of as the fraction of the distance from the subset model with predictors x_1 to the target model with predictors x_1 and x_2. In the initial plot shown in Figure 12.3, $\lambda = 0$, which indicates the subset model.

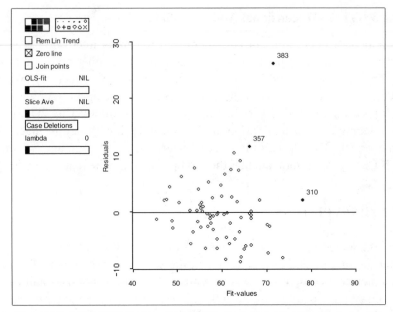

Figure 12.3. Initial *ARES* plot for adding *LG18* in the Berkeley Guidance Study for girls.

Pushing the cursor in the slide bar will increase λ toward its maximum value of 1. When $\lambda = 1$, the plot consists of {fitted values, residuals} computed from the fit of the target model, as shown in Figure 12.4. A generic plot in the sequence can be represented as residuals versus fitted values $\{\hat{y}(\lambda), e(\lambda)\}$ depending on the value of λ. As λ is changed, the plot is rapidly updated and the points appear to move continuously. The movement gives the information about relative merits of the target and subset models.

Repeat the animation several times, moving the slider from the right to the left and back again. Moving the slider from the right to the left corresponds to *removing LG18* from the target model, just as moving it from the left to the right corresponds to *adding LG18* to the subset model. Study the movement of points while repeating the animation. What are your general impressions? Most of the residuals move closer to zero. The movement is not parallel to the vertical axis since the fitted values are changing along with the residuals. The overall impression of decreasing residual magnitude provides a visualization of the usual t-test of the hypothesis that the coefficient of *LG18* is zero, given that the other predictors are in the target model. The *ARES* plot indicates that *LG18* is providing useful additional predictive power.

Study the behavior of the three marked points, with identification numbers 310, 357, and 383. Case number 383 is very poorly fit under both

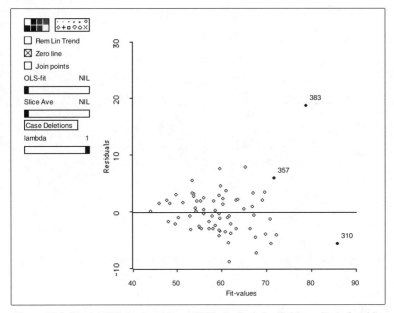

Figure 12.4. Final *ARES* plot for adding *LG18* in the Berkeley Guidance Study for girls.

models. Case number 310 is reasonably well fit under both models, but the fitted value changes enough for this case to reverse the sign of its residual. Case number 357 is poorly fit initially, but better fit when *LG18* is added to the model. Some points stay relatively stationary in the animation; others actually have residuals that increase as the variable is added.

Draw the *ARES* plot for *HT18*. How does your impression compare to the *ARES* plot for *LG18*? A recognizable decrease in the overall magnitude of the residuals is again apparent, but not nearly as great as for *LG18*. The *t*-values for *HT18* and *LG18* are 2.01 and 7.41, respectively. Does your visual impression reflect these *t*-values? Study the *ARES* plot for *ST18*. How does your impression compare to the other two? The points bounce around a bit, but nothing really changes. Does this impression agree with the *t*-value for *ST18*? In all three plots, also compare the behavior of the three cases marked in Figure 12.3.

Change now to another data set. Load the data in file `rat.lsp` in the `R-data` folder. These data are from an experiment on 19 rats that were given a dose of a drug, the dose being roughly in proportion to the body weight of the animal. At the end of the experiment, each rat was sacrificed, its liver was weighed, and the amount of drug in the liver was determined. The experimenters believed that the amount recovered, *y*, should be independent of the three predictors, *Dose*, *BodyWt*, and *LiverWt*; that is, the experimenters expected to see 0D structure. Set up the regression with *y*

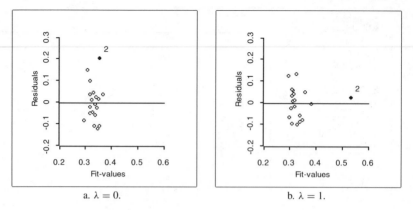

Figure 12.5. *ARES* plot for *Dose* in the rat data.

as the response and the other three variables as predictors. Examination of the printed output in the text window gives an unexpected finding: two of the three predictors have moderately large coefficients relative to their standard errors, and the overall F-test for all coefficients has a small p-value, suggesting that the coefficients may not all be zero.

Look at the *ARES* plot for the predictor *Dose*. The plots for $\lambda = 0$ and $\lambda = 1$ are shown in Figure 12.5. The *ARES* plot for *Dose* has two distinct features. The configuration of the main point cloud is fairly stable as λ changes. The exception to this is case 2, which brings us to the second feature. Case 2 starts in the subset model with the largest residual and ends in the target model with a residual near zero. The relative change in its fitted value is equally large. The *ARES* plot provides a visualization of the effects of a highly influential case, as will be discussed in Chapter 13. In effect, *Dose* is needed to model case 2 only. The t-value for *Dose* in the target model is 2.74, but it is only 0.4 after case 2 is deleted. The *ARES* plot for *Dose* shows no notable reduction in the magnitude of the residuals after deleting case 2.

12.2.2 *ARES* **Plots for Groups of Predictors**

The examples of *ARES* plots so far are all for adding a single predictor, but this is not a restriction on the methodology. Constructing an *ARES* plot for groups of predictors is easy.

Return to the Berkeley Guidance Study, with *WT18* as the response. Again select "ARES Plot" from the model's menu. Construct an *ARES* plot for the age 18 predictors by double clicking on them to move them to the right list. The *ARES* dialog has two options: the default "Sequential" and "As a group." Click on the "As a group" option and then push "OK."

The *ARES* plot is now viewed and interpreted just as an *ARES* plot for a single predictor; λ is still the fraction of the distance between the subset and target models. This *ARES* plot is a visualization of the F-test for comparing the subset model and the target model. View the two *ARES* plots for the age 18 predictors and for the age 9 predictors. What differences do you notice? Do the visualizations agree with the indications of the corresponding F-tests?

12.2.3 Sequential *ARES* Plots

In an *ARES* plot with sequential addition, the predictors in the group will be added one at a time in the order they were chosen in the dialog. The number to the right of the slide bar is in the structural form "$k.\lambda$." For example, the number "2.5" indicates that two predictors have already been added and the third predictor is currently being added with $\lambda = 0.5$.

If you choose to add all the predictors, the initial subset model consists only of an intercept. The sequential *ARES* plot is then a visualization of the sequential analysis-of-variance (ANOVA) table that results from the regression menu item "Sequential ANOVA" provided that the ordering of the variables is the same.

12.3 COMPLEMENTS

12.3.1 Added-Variable Plots

The added-variable plot first appeared in Cox (1958, p. 58); the name was coined by Cook and Weisberg (1982). Atkinson (1985) uses added-variable plots in many diagnostic methods. The 3D added-variable plot was proposed by Cook and Weisberg (1989). McCulloch (1993) gives an extension to four dimensions.

If the predictors are not linear, the residuals $e(x_2|x_1)$ do not appropriately adjust x_2 for its relationship with x_1, and an added-variable plot will overstate the contribution of x_2.

Removing the linear trend from an added-variable plot replaces the vertical axis of a plot by the residuals e, and hence they become residual plots. If all the predictors are linear predictors, then detrended added-variable plots, in 2D and 3D, are the same as *ALP* residual plots.

12.3.2 Constructing *ARES* Plots

A frame of an *ARES* plot is of the form $\{\hat{y}(\lambda), e(\lambda)\}$. *ARES* plots are easy to construct because

$$\hat{y}(\lambda) = \hat{y}(0) + \lambda(\hat{y}(1) - \hat{y}(0))$$

and

$$e(\lambda) = y - \hat{y}(\lambda) = e(0) - \lambda(\hat{y}(1) - \hat{y}(0))$$

where $\hat{y}(0)$ and $e(0)$ are the fitted values and residuals for the subset model and $\hat{y}(1)$ designates the fitted values for the target models.

These expressions provide an operational meaning to the interpretation of λ as the fractional distance between the subset and target models since $\hat{y}(\lambda)$ is indeed located between $\hat{y}(y|x_1)$ and $\hat{y}(y|x_1, x_2)$ as indicated by the fraction λ. The value $\lambda = 0.5$, for example, means that $\hat{y}(0.5)$ is the average of $\hat{y}(y|x_1, x_2)$ and $\hat{y}(y|x_1)$.

The *ARES* plot was proposed by Cook and Weisberg (1989). They are also discussed in Cook and Weisberg (1994) for generalized linear models. The rat data are from Weisberg (1985, p. 122).

EXERCISES

12.1. In this exercise, we verify the construction of the added-variable plot by constructing the plot for *HT9* in the Berkeley Guidance Study for girls. This plot is shown in Figure 12.1. Load the `BGSgirls.lsp` from the `R-data` folder, and construct the model as suggested in Section 12.1. Do not change the name of this model, `BGS-girls`. To get the added-variable plot, we need the residuals from the regression of *HT18* on *ST9* and *WT9* and also the residuals from the regression of *HT9* on *ST9* and *WT9*. Using the "New Model..." item in the regression menu twice, create these two regression models and name them *h18-model* and *h9-model*, respectively. The added-variable plot is just the residuals from one model versus the residuals from the other model. The two sets of residuals can be viewed by typing the following commands:

```
(send h18-model :residuals)
(send h9-model :residuals)
```

Next, type in the following command:

```
(rcode :data (list (send h18-model :residuals)
                   (send h9-model :residuals))
       :data-names (list "H18-residuals" "H9-residuals")
       :name "AVP-demo")
```

This will give a standard regression dialog. Choose *H18-residuals* as the response and *H9-residuals* as the single predictor, and push "OK."

The added-variable plot is given by the 2D plot

$$\{\textit{H9-residuals, H18-residuals}\}$$

Use the regression to verify numerically the following general facts about added-variable plots:

1. The intercept estimated from the added-variable plot is zero. This result requires that the intercept be included in the target model.

2. The slope in the added-variable plot is the same as the slope for *HT9* in the original regression.

3. The residuals from the regression of *H18-residuals* on *H9-residuals* are the same as the residuals from the fit of the target model.

4. Since the residuals are the same, estimates of variance are the same except that the regression of *H18-residuals* on *H9-residuals* ignores the adjustment needed for the other predictors and therefore gets the degrees of freedom wrong.

5. Suppose t_2 is the usual t-statistic for testing $\beta_2 = 0$ in the original regression model, where β_2 corresponds to the single predictor *HT9*. Similarly, let t^* be the t-statistic for testing the hypothesis that the slope in the regression of *H18-residuals* on *H9-residuals* is zero. These two statistics are slightly different. Show that $t_2 = [(n-p-1)/(n-2)]t^*$. The constant corrects the degrees of freedom in the estimates of variance.

12.2. Consider again the birth weight data in the file `birthwt.lsp` in the `R-data` folder. First set up the regression with *Term* as the response and the other three variables as predictors. Verify that the target model (2.1) is reasonable for these data. Use added-variable plots to assess the importance of each of the variables in the model. Use a sequential *ARES* plot, adding all three predictors. Describe what you see in these plots, and summarize results.

12.3. Use added-variable plots and *ARES* plots to study the predictor contributions to your final model for the land rent data discussed in Exercise 6.7.

CHAPTER 13

Influence and Outliers

In almost any plot of real data, a few points will be separated from the central trend. Our main focus has been on understanding the relationship between the response and the predictors that is appropriate for most of the data without in-depth study of separated points, but such points can merit special attention. They may cause visual problems with plots: Outlying points can cause us to miss important trends by reducing resolution. Outlying points can also have a large *influence* on conclusions. Were these points deleted, conclusions may be quite different. Cases corresponding to unusual points may present new and unexpected information when they are studied individually.

The main tool for examining the effects of individual points on an analysis is the "Case Deletions" plot control. This control allows us to discover how an analysis changes when points are deleted. This general idea was illustrated several times in previous chapters.

Much more can be said about the detection of influential cases and outliers when the target model (11.1) is used. In this chapter we give an overview of methods available in the *R-code* that can be used to understand the influence of unusual points on the fit of model (11.1). We include in the last section of this chapter a discussion of probability plots.

Linear predictors are not required in this chapter.

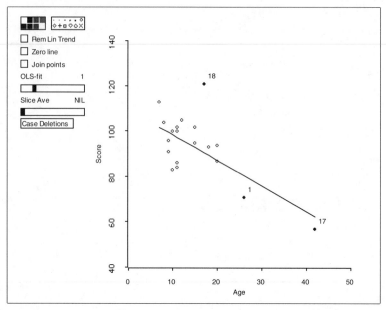

Figure 13.1. The adaptive score data.

13.1 INFLUENCE

Load the file `adscore.lsp` from the `R-data` folder. This is a small data set with observations on 21 children, giving their *AGE* in months at first spoken word, and a *SCORE*, which is a measure of the development of the child. The goal is to study the conditional distributions of *SCORE* given *AGE*. Construct the plot {*AGE, SCORE*} as shown in Figure 13.1. The ols linear fit and three separated points are labelled on the plot. *SCORE* appears to decrease with *AGE* and case 18 seems to be poorly fit by the linear trend, relative to the other data. Cases 1 and 17, while not poorly fit, are interesting by virtue of their relatively large values of *AGE*.

Suppose we fit a target model

$$y|x = \beta_0 + \beta^T x + \varepsilon \qquad (13.1)$$

with $\text{var}(y|x) = \sigma^2$ to these data. How do you think the fit of the target model will change when cases marked 1 and 17 are deleted? What will happen to the estimates of β_0, β, and σ^2? Select those two points and use the "Case Deletions" pop-up menu to delete them. This will not change the remaining points, but the fitted line does change. Use the "Display fit" item in the regression menu to see the new fitted model.

Table 13.1. Adaptive Score Data Estimates with Cases Removed

	Intercept	Slope	SE(slope)	$\hat{\sigma}$	R^2
All Data	110	-1.1	.31	11.0	.41
Not 1, 17	98	-0.1	.62	10.5	.00
Not 1	109	-1.0	.35	11.1	.35
Not 17	106	-0.8	.52	11.1	.11
Not 18	109	-1.2	.57	8.6	.57

Selected summaries from regressions with a few cases deleted are given in Table 13.1. Without cases 1 and 17, the relationship between *AGE* and *SCORE* almost completely disappears: A small fraction of the data effectively determines the fitted model. Such cases are called *influential*. In this example, the two cases with large values of *AGE* are influencing the fit.

When we have more than two predictors, finding influential points may not be so easy. However, given a target model like (13.1), we can compute a diagnostic statistic called *Cook's distance* that summarizes some of the information about the influence of each case. Suppose we wanted to know if the ith case is influential for estimating β. We could imagine computing $\hat{\beta}_{(i)}$, the estimate of β computed without case i, and then computing the difference between the estimates. If the difference $\hat{\beta}_{(i)} - \hat{\beta}$ is large, then we could declare case i to be influential because its deletion gives a new estimate of β that is very different from the estimate that uses all the data. Cook's distance combines this vector difference into a summary number D that is a squared distance between $\hat{\beta}_{(i)}$ and $\hat{\beta}$. A useful property of D is that it can be computed without refitting the regression. Cases for which D is the largest are candidates for influential cases. A formula for computing D is given in Section 13.5.1.

Return to the adaptive score data and restore any deleted cases. Draw a plot of {case numbers, Cook's distances}. Case 17 has the largest value of $D = 0.68$. Deletion of case 17 will cause the largest change in estimated coefficients, in agreement with what we found graphically. Case 1 does not have a particularly large value of D when all the data are used. However, try deleting case 17 from the regression and see how D for case 1 changes. You will need to rescale the plot {case numbers, Cook's distances} by using the item "Rescale Plot" from the plot's menu.

13.2   MEAN-SHIFT OUTLIERS

We can expand the target model (13.1) to allow for a specific type of outlier, and this will lead to an outlier test. Suppose the target model holds for all

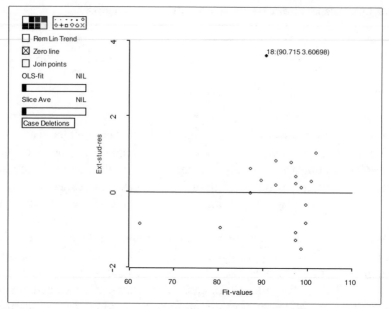

Figure 13.2. Externally Studentized residuals for the adaptive score data. The "Show Coordinates" mouse mode was used to print the coordinates of point 18 on the plot.

cases except the ith case, and for this one case $E(y|x_i) = \beta_0 + \beta^T x_i + \delta$, so the regression function for the ith case is shifted by an amount δ. If $\delta = 0$, then the target model holds for all the data. A test of $\delta = 0$ is therefore a test for the ith case to be a *mean-shift outlier*.

A single model for all the cases can be written as

$$y|(x, u) = \beta_0 + \beta^T x + \delta u + \varepsilon \qquad (13.2)$$

where $u = 1$ for the ith case and $u = 0$ for all others. This is just a linear model with one additional predictor u, so a test of $\delta = 0$ can be obtained as a usual t-statistic for $\delta = 0$ in the fit of (13.2). This statistic will have $n - p - 2$ degrees of freedom and is equal to the ith *externally Studentized residual*.

The plot of {fitted values, externally Studentized residuals} given in Figure 13.2 displays all n externally Studentized residuals. Unless the location of an outlier is known in advance, outlier testing is usually based on the largest in absolute value of the externally Studentized residuals; in Figure 13.2, the largest value is about 3.61 for case 18. To get a significance level for the test, we need to adjust for the multiple testing that is inherent in using the largest of n statistics. Using the *Bonferroni inequality*, an upper bound on the p-value can be obtained by multiplying the standard

$t(n-p-2)$ significance level by n. The function `outlier-pvalue` can be used to calculate this value:

```
> (outlier-pvalue 3.60698 18 21)
0.0420453
```

suggesting that case 18 may be a mean-shift outlier.

Often outliers will be influential, but this need not be so. For example, in Table 13.1 we see that deleting the suspected outlier case 18 has almost no effect on the coefficient estimates, but its deletion reduces the estimate of σ^2 from 11 to 8.6. This case is influential for estimating σ^2, but not for estimating coefficients.

13.3   EXAMPLE: FUEL DATA

Load the file `fuel90.lsp` from the `R-data` folder. These data consist of several measurements on the U.S. states and the District of Columbia. Treat the per person fuel consumption (*FUEL/POP*) as the response and use income (*INC*), vehicles per person (*VEH/POP*), *TAX* rate, and average miles per vehicle (*VM/VEH*) as predictors. Examine the scatterplot matrix of the predictors and the response. Because of the relatively narrow range for each of the predictors, transformations will not be very effective in improving linearity between them. Fortunately the relationships between the predictors appear to be fairly linear, although there are a few separated points in most of the plots. The inverse partial response plots are difficult to interpret because of one outlying point for Wyoming (WY), a large state with a small population. Apart from this one point, the inverse partial response plots for *INC* and *VEH/POP* are reasonably linear, while the remaining two plots show no trends at all. Taken together, these plots give no information to contradict 1D structure.

Focus on the partial response plot for *VEH/POP*. Five points appear separated from the main body of the plot, as shown in Figure 13.3. Three of these states have the lowest per capita fuel consumption, one has the highest fuel consumption, and one state has high fuel consumption given its value of *VEH/POP*. Mark these points with a color or symbol for future reference.

Next, select the item "AVP–All 2D" from the regression menu. As you cycle through the added-variable plots, note the locations of the points you have marked. The most outlying single point is consistently for Wyoming. In most of the added-variable plots, Wyoming is separated from the bulk

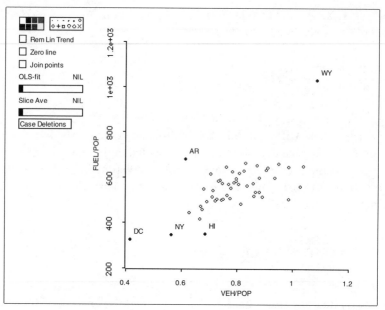

Figure 13.3. Partial response plot for *VEH/POP* for the fuel data.

of the data both vertically, suggesting this state might be an outlier, and horizontally, suggesting that this state has values for the predictors that are different from most states. Cases separated horizontally in added-variable plots are *leverage* cases. Combining vertical and horizontal separation, Wyoming is probably influential, and its deletion may change conclusions. Delete Wyoming and see what happens to the added-variable plots. The most striking change is for *INC*, which changes from a plot with little evidence of a linear trend to one that has a negative trend. This is confirmed by using the "Display fit" item in the regresison menu and comparing coefficient estimates with and without Wyoming. Virtually all the coefficients show large changes when Wyoming is deleted. Since only one of the cases is influential, Cook's distance would be very effective in locating this case, as its value is about 1.33, while the second largest value is only 0.18.

13.4 PROBABILITY PLOTS

A 2D probability plot is used to study the distribution of a random sample. Suppose we have n observations $v_1, \ldots, v_n$, and we want to examine the hypothesis that these are a sample from a specific distribution, such as a normal distribution. Let $v_{[1]} \leq \cdots \leq v_{[n]}$ be the v_i's reordered from smallest to largest. Let $u_{[1]} \leq \cdots \leq u_{[n]}$ be the expected values of

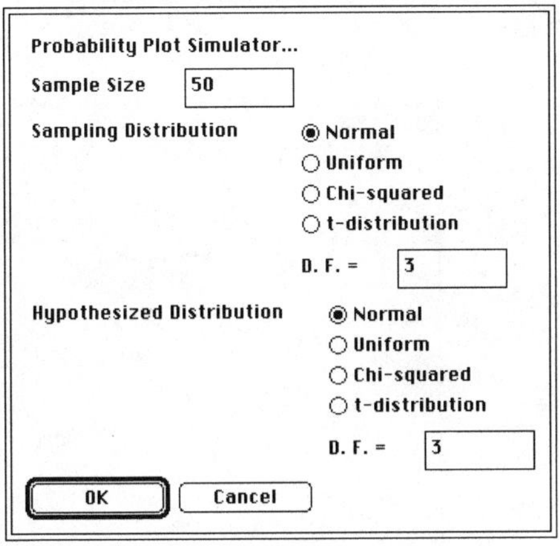

```
┌─────────────────────────────────────────────┐
│  Probability Plot Simulator...               │
│                                              │
│  Sample Size    ┌─────────────┐              │
│                 │ 50          │              │
│                 └─────────────┘              │
│                                              │
│  Sampling Distribution    ⦿ Normal           │
│                           ○ Uniform          │
│                           ○ Chi-squared      │
│                           ○ t-distribution   │
│                                              │
│                           D. F. = ┌────────┐ │
│                                   │ 3      │ │
│                                   └────────┘ │
│  Hypothesized Distribution  ⦿ Normal         │
│                             ○ Uniform        │
│                             ○ Chi-squared    │
│                             ○ t-distribution │
│                                              │
│                           D. F. = ┌────────┐ │
│                                   │ 3      │ │
│                                   └────────┘ │
│  ┌──────────┐  ┌──────────┐                  │
│  │   OK     │  │  Cancel  │                  │
│  └──────────┘  └──────────┘                  │
└─────────────────────────────────────────────┘
```

Figure 13.4. Dialog for the probability plot demonstration.

an ordered sample of size n from the hypothesized distribution. That is, $u_{[i]} = E(z_{[i]})$ where $z_1, \ldots, z_n$ is a sample from the hypothesized distribution.

If the true sampling distribution of the v's is in fact the same as the hypothesized distribution, we would expect the ordered v's to be linearly related to the u's. A *probability plot* is given by $\{u_{[i]}, v_{[i]}\}$. If the sampling distribution of the v's is the same as the hypothesized distribution, then the plot should be approximately straight; if the plot is curved, then we have evidence that the v's are not from the hypothesized distribution.

Judging if a probability plot is straight requires practice. To help gain experience, load the demonstration file `demo-inf.lsp` from the `R-data` folder, and select the item "Probability Plots" from the "Demo:Inf" menu. You will get the dialog shown in Figure 13.4. You can specify the sample size, the true sampling distribution the computer will use to generate the data, and the hypothesized distribution the computer will use to compute the $u_{[i]}$.

For the first try, set both distributions to be normal, leave the sample size at 50, and then push "OK." A probability plot similar to Figure 13.5 will result. The points in this plot should be approximately straight because the sampling and hypothesized distributions are the same. A new sample from the sampling distribution can be obtained by pushing on the "New Sample" button on the plot; this can be repeated many times.

After looking at several plots with sample size 50, start the demonstration over by again selecting "Probability Plots" from the menu, but this time set the sample size to 10. Can you judge normality as easily for a sample of size 10 as you could for a sample of size 50?

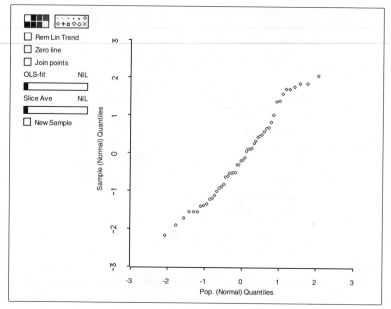

Figure 13.5. A sample probability plot.

Nonnull shapes of probability plots can be studied by setting the sampling distribution and the hypothesized distribution to be different in the dialog; this will require selecting the "Probability Plots" item again. Relative to the normal, the uniform distribution has short tails, or few large values. The t distributions, for which you must specify the degrees of freedom, have long tails for small degrees of freedom. The χ^2 distributions are skewed; the smaller the degrees of freedom, the greater the skewness.

Probability plots are obtained in the *R-code* by selecting the item "Probability Plot of..." from the regression menu. Selecting this item will produce the dialog shown in Figure 13.6. From this dialog, click on the quantity to be plotted and the target distribution to get the $u_{[i]}$. If the distribution chosen is χ^2 or t, then the number of degrees of freedom must be specified as well.

13.5 COMPLEMENTS

13.5.1 Residuals and Leverages

Many books on regression suggest using rescaled versions of residuals in plots. Given a target model like (13.1), the residuals have unequal variances,

$$\text{var}(e|x) = \sigma^2(1 - h) \tag{13.3}$$

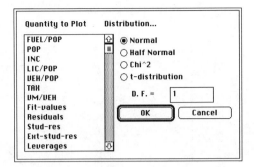

Figure 13.6. Probability plot dialog.

where h is usually called the leverage. The value of the leverage is always between 0 and 1 and depends only on the predictors in the model, not the response. The leverage is largest for cases with x farthest from $\bar{x}$ and smallest for cases with x close to $\bar{x}$. The sum of the h is equal to $p+1$, the number of coefficients in model (13.1) including the intercept, so the average value of h is $(p+1)/n$. If n is large compared to p, then the average leverage $(p+1)/n$ is small and the individual leverages will usually deviate little from the average.

Residuals can be *Studentized* to correct for unequal variance. This is usually done in one of two ways, depending on the method of estimating σ^2. The *Studentized residuals* are given by

$$ r = \frac{e}{\hat{\sigma}\sqrt{1-h}} \qquad (13.4) $$

where $\hat{\sigma}$ is the square root of the usual estimate of σ^2 computed using all n cases in the computations. If $\hat{\sigma}$ is replaced in (13.4) with $\hat{\sigma}_{(i)}$, the usual estimate found without using case i in the computations, the resulting residual is an externally Studentized residual. If the target model holds, then both types of Studentized residuals have mean zero and constant variance. For the target model (13.1), the two types of Studentized residuals are related by a simple nonlinear equation, so they are practically equivalent.

Cook's distance is a function of the Studentized residual, the leverage, and the number of predictors. The explicit formula is

$$ D = \frac{r^2}{p+1} \frac{h}{1-h} $$

13.5.2   Local Influence

The use of added-variable plots to identify influential cases described at the end of Section 13.3 is based in part on a generalized method of influence

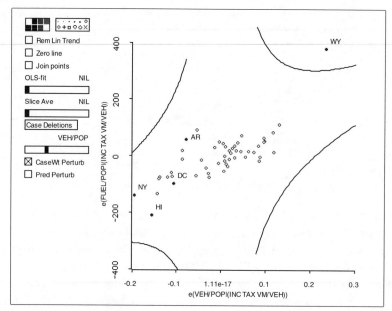

Figure 13.7. Case weight perturbation contours for the coefficient of *VEH/POP* in the fuel data.

assessment. The basic idea behind influence analysis is that a regression solution should be stable: Small changes in the data should not produce large changes in the results. Deleting cases is one way of introducing small changes in the data, but there are others as well.

We might assume that the variance function is not quite constant. Suppose we set $\text{var}(\varepsilon|x) = \sigma^2/\omega$, where ω is positive. We can study the change in a single coefficient estimate, say $\hat{\beta}_1$, as ω changes. Let $\hat{\beta}_1(\omega)$ denote the *weighted least squares* estimate of β_1 with weights ω. If $\omega = 1$ for all cases, then $\hat{\beta}_1(\omega) = \hat{\beta}_1$, the ols estimate, but for other values of ω we can get a different value. We can get a worst case by solving the following problem: Find ω not too far from 1 for every case such that $\hat{\beta}_1(\omega)$ is as far from $\hat{\beta}_1$ as possible. Cases whose values for ω are most different from 1 are potentially influential cases. Finding the worst case is the same as the mathematical problem of maximizing the rate of change in $\hat{\beta}_1(\omega)$ as ω is varied. This maximization problem can be solved, and the potentially influential cases can be identified in the added-variable plot for x_1.

Construct the added-variable plot for *VEH/POP* in the fuel data. In the plot controls push the button "CaseWt Perturb." A contour with four segments will appear on the plot, as shown in Figure 13.7. This is a typical contour showing relative impact when case weights are perturbed. All cases falling on the contour lines will have the same local influence. The farther the contour from the center of the plot, the greater the influence. From the plot, modifying the case weight for Wyoming will result in the greatest rate of change in the estimate of the coefficient of *VEH/POP*.

The plot control "Pred Perturb" is used just like the previous control except it is the values of the predictor rather that the case weight that are perturbed. Pushing this control on the plot in Figure 13.7 will show which cases have the greatest impact on the ols estimate of the coefficient of *VEH/POP* when the corresponding values of *VEH/POP* are changed. The contours in this case will always be straight lines.

13.5.3 Simulated Envelopes

A visual aid can be constructed via simulation to help judge whether or not a probability plot is straight. Assume the target model (13.1) holds.

1. Let $\hat{y}$ be the fitted values from the target model, and let $\hat{\sigma}$ be the estimate of σ. Compute a simulated response $y_s = \hat{y} + \hat{\sigma} z$, where z is an $n \times 1$ vector of standard normal random deviates.

2. Fit the target model, but with y_s as the response, and then compute the probability plot for the residuals.

3. Repeat steps 1 and 2 *m* times; the *R-code* uses $m = 19$.

4. We now have *m* simulated residuals for each plotting location on the horizontal axis of a probability plot. Add the smallest and the largest of these *m* values to the plot at each plotting location on the horizontal axis.

5. Join all the smallest values and all the largest values. This will give a simulated *envelope*. If the assumed population distribution is correct, one would expect that all the points in the observed probability plot will fall within the envelope a fraction $m/(m + 1)$ of the time; if $m = 19$, this is 95% of the time.

In the *R-code* a simulated envelope is obtained by pushing the "Simulated Env" button on a probability plot and waiting for the lengthy calculation to be completed. If the observed plot falls within the simulated envelope, then the assumed null distribution for the residuals is supported; otherwise, it is not supported.

For the target model (11.6) fit to the *Casuarina* data discussed in Section 11.6, the probability plot of Studentized residuals is shown as Figure 13.8. Although the plot is not particularly straight, it does fall entirely within

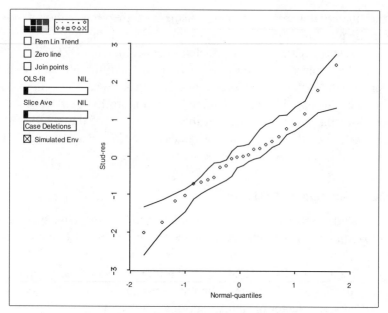

Figure 13.8. Probability plot for the *Casuarina* data.

the simulated envelope, so a hypothesis of normality of the errors is not contradicted.

The simulated envelope assumes that the model used is correct, and all that is at issue is the distribution of the errors. In realistic problems, normality or any other modeling assumptions such as linearity, 1D structure, and constant variance may in fact not hold, and many incorrect assumptions can result in a curved probability plot. In particular, outliers will often appear in probability plots as isolated points in the left or right tail of the plot. To illustrate this, construct the probability plot of the Studentized residuals in the adaptive score data. Case 18 falls well above the main trend in the plot.

13.5.4 References

A good elementary introduction to residuals, leverages, outliers, and influence measures is given by Fox (1991). More advanced treatments are given by Cook and Weisberg (1982) and by Atkinson (1985). Leverages are discussed in Hoaglin and Welsch (1980). Deletion influence measures were introduced by Cook (1977). The use of added-variable plots for studying local influence is developed in Cook (1986a). Many other influence measures have been suggested; see Cook, Peña, and Weisberg (1988) and Cook (1986b) for comparisons.

The probability plot described here is often called a *QQ-plot*, as it is a plot of the quantiles of the hypothesized distribution against the observed quantiles of the sample. Gnanadesikan (1977) is a standard reference on these plots. The simulated envelopes for probability plots were suggested by Atkinson (1981).

The adaptive score data first appeared in Mickey, Dunn, and Clark (1967). The muscles data used in Exercise 13.6 are from Cochran and Cox (1957, p. 176).

EXERCISES

13.1. Complete the analysis of the mussels data, paying careful attention to the effect of individual cases on the regression. This analysis was begun in Exercise 10.2.

13.2. Use the adaptive score data to verify certain properties of the residuals and fitted values. First, verify that the residuals add to zero, apart from rounding error. This can be done using the :residuals method and typing

```
(sum (send adscore :residuals))
```

Similarly, using the :leverages method, verify that the sum of the leverages is equal to the number of coefficients in the model, including the intercept, which is 2 for these data. Next, verify that $y = e + \hat{y}$ using the :y, :residuals and :fit-values commands by typing

```
(plot-points (send adscore :y)
   (+ (send adscore :residuals) (send adscore :fit-values)))
```

Next, verify that the regression of e on $\hat{y}$ has slope equal to zero. Finally, show that the square of the correlation between y and e is equal to $1 - R^2$. These results hold for any ols fit of a linear model with an intercept.

13.3. Compute confidence curves for the Box–Cox power transformation of the response for the fuel data discussed in Section 13.3. Delete Wyoming and recompute the confidence curves. Does Wyoming influence the choice of a response transformation? Could the influence of Wyoming have been anticipated using the inverse fitted-value plot $\{y, \hat{y}\}$ described in Section 10.2?

13.4. Analyze the Berkeley Guidance study for boys using the same variables as were used in Section 7.5 for girls. The data are in the file BGSboys.lsp in the R-data folder.

13.5. Examine the Big Mac data for influential cases and for outliers. Which cities seem to be different from the rest? How are they different? Do they strongly influence any conclusions?

13.6. The data in file muscles.lsp in the R-data folder comprise two replications of a $4 \times 4 \times 3$ factorial experiment on rats to investigate the use of electrical stimulation to prevent deterioration of denervated muscles. The response y is the weight (1 unit = 0.01g) of the denervated muscle at the end of the experiment. Since larger animals tend to have larger muscles, the weight x of the untreated bilateral muscle was used as a covariate. The other factors in the experiment were *Rep*, the replication number, either 0 or 1; *TrtTime*, the length of stimulation in minutes, 1, 2, 3, or 5; *Trt/day*, the number of treatments per day, 1, 3, or 6; and *Trt*, a qualitative factor for the type of current used, 1 = galvanic, 2 = faradic, 3 = 60 cycle, or 4 = 25 cycle.

Provide a complete analysis of these data, and summarize your findings. Once you obtain a target model, be sure to analyze the data for outliers and influential points.

CHAPTER 14

Confidence Regions

This chapter is about confidence regions for two or three coefficients in a linear model, including their implementation in the *R-code*. Linear predictors are not required in this chapter.

14.1 CONFIDENCE REGIONS IN THE *R-code*

Load the file BGSboys.lsp in the R-data folder. This file contains data on boys from the Berkeley Guidance Study, similar to the data on girls used elsewhere in this book. We assume that weight at age 18, *WT18*, depends on measurements of height, weight, leg circumference, and strength at age 9, via the model

$$WT18 = \beta_0 + \beta_1 HT9 + \beta_2 WT9 + \beta_3 LG9 + \beta_4 ST9 + \varepsilon \qquad (14.1)$$

with constant variance σ^2 and normally distributed errors.

To construct a joint 95% confidence region for β_2 and β_4 in model (14.1), select "Confidence Regions" from the regression menu. In the resulting dialog, double click on *WT9* and *ST9* to move them from the left list to the right list and push "OK." The result is shown in Figure 14.1a. The cross hairs mark off marginal 95% confidence intervals on the coordinate axes for β_2 and β_4. For example, the 95% confidence interval for β_2 runs from about -0.6 to 1.4.

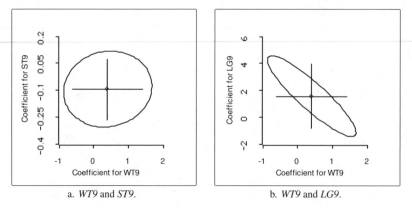

a. *WT9* and *ST9*.   b. *WT9* and *LG9*.

Figure 14.1. Joint and marginal 95% confidence regions of the Berkeley Guidance Study for boys.

The elliptical region in Figure 14.1a is a joint 95% confidence region for β_2 and β_4. Confidence regions for two coefficients in a normal linear regression model are always elliptical. The ellipse in Figure 14.1a is nearly a circle, but this characteristic of the shape depends on the aspect ratio in the plot. The cross hairs in Figure 14.1a sit well within the joint confidence region so the range allowed for either coefficient in the joint region is larger than the corresponding range of either marginal confidence interval.

Construct a joint confidence region for the coefficients for *WT9* and *LG9*, as shown in Figure 14.1b. Figures 14.1a and b are qualitatively different. The major axis of the ellipse in Figure 14.1b has a negative slope, so $\hat{\beta}_2$ and $\hat{\beta}_3$ are negatively correlated. When the scaling in the plot is the default scaling as in these figures, elongation of the ellipse indicates large correlation. The cross hairs in Figure 14.1b extend outside of the joint confidence region because of the high correlation. In Figure 14.1a the near circularity suggests that the correlation between the coefficients of *WT9* and *ST9* is small.

14.2   CONFIDENCE REGIONS IN TWO-PREDICTOR MODELS

Let's now consider a joint confidence region for the regression coefficients (β_1, β_2) in a two-predictor model,

$$y|x = \beta_0 + \beta_1 x_1 + \beta_2 x_2 + \varepsilon$$

with independent normal errors having mean 0 and constant variance σ^2. The estimated 2×2 covariance matrix $\widehat{\text{var}}(\hat{\beta})$ of the ols estimates $\hat{\beta}^T = (\hat{\beta}_1, \hat{\beta}_2)$ is

$$\widehat{\text{var}}(\hat{\beta}) = \hat{\sigma}^2 \begin{pmatrix} S_{11} & S_{12} \\ S_{12} & S_{22} \end{pmatrix}^{-1}$$

where

$$S_{11} = \sum_{i=1}^{n}(x_{i1} - \bar{x}_1)^2 \qquad S_{22} = \sum_{i=1}^{n}(x_{i2} - \bar{x}_2)^2$$

and

$$S_{12} = \sum_{i=1}^{n}(x_{i1} - \bar{x}_1)(x_{i2} - \bar{x}_2)$$

A joint $(1-\alpha) \times 100\%$ confidence region for (β_1, β_2) is the set of all values of the 2×1 vector β that satisfies the inequality

$$(\beta - \hat{\beta})^T[\widehat{\text{var}}(\hat{\beta})]^{-1}(\beta - \hat{\beta}) \leq 2F(1 - \alpha, 2, n - 3) \qquad (14.2)$$

where $F(1 - \alpha, 2, n - 3)$ is the percentage point of the F-distribution with 2 and $n - 3$ degrees of freedom that leaves an area of α under the right tail. The points satisfying this inequality fall inside an ellipse. The center of the ellipse is at $\hat{\beta}$. The size of the ellipse is partially controlled by the choice of the level $1 - \alpha$. As we require more confidence, $1 - \alpha$ becomes bigger, $F(1 - \alpha, 2, n - 3)$ becomes bigger, and the area of the ellipse increases.

After a bit of algebra, the element of $\widehat{\text{var}}(\hat{\beta})$ that corresponds to $\widehat{\text{var}}(\hat{\beta}_1)$ is

$$\widehat{\text{var}}(\hat{\beta}_1) = \frac{\hat{\sigma}^2}{S_{11}(1 - r_{12}^2)} \qquad (14.3)$$

where $r_{12} = S_{12}/(S_{11}S_{22})^{1/2}$ is the sample correlation coefficient between x_1 and x_2. The standard errors of the coefficient estimates determine the lengths of marginal intervals. Expression (14.3) tells us that the length of a confidence interval increases as r_{12}^2 increases, so high correlation between the predictors gives relatively long confidence intervals.

The shape of a joint confidence interval depends on $\text{corr}(\hat{\beta}_1, \hat{\beta}_2)$, the correlation between $\hat{\beta}_1$ and $\hat{\beta}_2$, and on scale factors. If all scale factors are fixed, the joint confidence region becomes elongated as $|\text{corr}(\hat{\beta}_1, \hat{\beta}_2)|$ increases. What will make $|\text{corr}(\hat{\beta}_1, \hat{\beta}_2)|$ large? There is a close connection between this correlation and r_{12}:

$$\text{corr}(\hat{\beta}_1, \hat{\beta}_2) = -r_{12} \qquad (14.4)$$

The correlation between the ols estimates of β_1 and β_2 in a regression model with just two predictors is the negative of the sample correlation between the predictors.

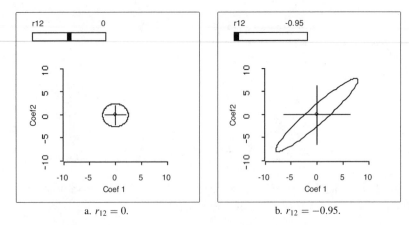

a. $r_{12} = 0$.   b. $r_{12} = -0.95$.

Figure 14.2. Confidence region demonstration.

Load the demonstration `demo-cr.lsp` from the `R-data` directory. A brief description of the demonstration will be printed and a 2D plot with a slider labelled "r12" will appear on the computer screen: "r12" is the sample correlation r_{12} between x_1 and x_2.

This demonstration is for a regression problem with two predictors scaled to have mean 0 and $S_{11} = S_{22} = 1$. In this scaling and with an aspect ratio of 1, the shape of a confidence region depends only on the correlation r_{12}. The initial plot shown in Figure 14.2a is the joint 95% confidence region for (β_1, β_2) when $r_{12} = 0$. The cross hairs on the plot give marginal intervals, assuming that $\hat{\beta}^T = (0, 0)$ and that $\hat{\sigma}^2 = 1$. The slide bar controls the sample correlation r_{12} between x_1 and x_2. As the slide bar is moved, the confidence regions are redrawn.

Figure 14.2a with $r_{12} = 0$ is similar to Figure 14.1a. Before moving the slider, consider what will happen as it is moved to the right. What will happen to the cross hairs? The demonstration has been constructed to keep the center at the origin, so only lengths and orientation can change. What will happen to the ellipse? How will its orientation change?

As the slider is moved to the right, the marginal intervals depicted by the cross hairs get longer. Was this predicted by the previous discussion? Also, the slope of the major axis of the ellipse is now negative when r_{12} is positive. Was this predicted by the previous discussion? The plot when the sample correlation $r_{12} = -0.95$ is shown in Figure 14.2b.

14.3   CONFIDENCE REGIONS IN MODELS WITH MANY PREDICTORS

With minor modifications, we can apply our understanding of confidence regions for two-predictor regression to the more general model,

$$y|z = \beta_0 + \beta_1 z_1 + \beta_2 z_2 + \beta_3^T z_3 + \varepsilon \qquad (14.5)$$

where z_1 and z_2 are single predictors and all remaining predictors are collected in the vector z_3. We have changed the generic designation of a predictor from x to z to help avoid confusing these predictors with those of the two-predictor case in Section 14.2.

How do we construct a joint confidence region for (β_1, β_2)? Aside from the presence of z_3, this is the same problem considered in Section 14.2. The required confidence region can be constructed by first getting $e(z_1|z_3)$, the residuals from the regression of z_1 on z_3, and $e(z_2|z_3)$, the residuals from the regression of z_2 on z_3. Both of these regressions should include an intercept term because there is an intercept term in (14.5). Now go back to Section 14.2 and replace x_1 with $e(z_1|z_3)$ and x_2 with $e(z_2|z_3)$. Except for a change in degrees of freedom, the discussion of the previous section now applies verbatim to a confidence region for (β_1, β_2) in (14.5). To apply equation (14.2), replace $F(1 - \alpha, 2, n - 3)$ with $F(1 - \alpha, 2, n - p - 1)$.

The shape of the confidence region in two-predictor regression is determined by the sample correlation r_{12}. In the many-predictor generalization, this is replaced by the sample correlation between $e(z_1|z_3)$ and $e(z_2|z_3)$. This correlation is called the sample *partial correlation* between z_1 and z_2 adjusted for z_3. It gives the sample correlation between z_1 and z_2 after removing any linear association with z_3. The sample partial correlation between z_1 and z_2 controls the behavior of joint confidence regions in multiple regression problems.

The residuals $e(z_1|z_3)$ and $e(z_2|z_3)$ are two of the variables needed to get a 3D added-variable plot for adding z_1 and z_2 after accounting for the effects of z_3. We thus have another role for 3D added-variable plots, but now linear predictors are not required. If we begin with the 3D added-variable plot $\{e(z_1|z_3), e(y|z_3), e(z_2|z_3)\}$ in the "Home" position and then use the "Pitch" control to rotate to the 2D plot $\{e(z_1|z_3), e(z_2|z_3)\}$, we can gain a visual impression of the size of the sample partial correlation. This in turn may provide a qualitative impression about the joint confidence region. In particular, if replacing the out-of-page variable O by $e(O|H)$ is needed for adequate resolution of a 3D added-variable plot, then the corresponding joint confidence region will probably be long and thin.

Figure 14.3a gives a view of the 3D added-variable plot for *WT9* and *ST9* in the Berkeley Guidance Study, rotated to display the static 2D plot $\{e(z_1|z_3), e(z_2|z_3)\}$. This figure provides a visual impression of the sample partial correlation that enters into the construction of Figure 14.1a. Similarly, Figure 14.3b gives a visual impression of the large positive sample partial correlation for Figure 14.1b.

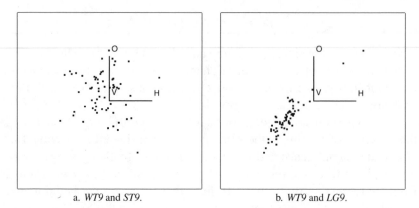

<table>
<tr><td>a. WT9 and ST9.</td><td>b. WT9 and LG9.</td></tr>
</table>

Figure 14.3. Three-dimensional added-variable plots for the Berkeley Guidance Study for boys, rotated to display sample partial correlations.

14.4 GENERAL CONFIDENCE REGIONS

Joint confidence regions for more than two coefficients follow the same general ideas. In particular, letting β_S denote an arbitrary subset of the coefficients in a multiple regression model, a $(1 - \alpha) \times 100\%$ confidence region for β_S is the set of all values of β_S that satisfy the following inequality:

$$(\beta_S - \hat{\beta}_S)^T [\widehat{\text{var}}(\hat{\beta}_S)]^{-1} (\beta_S - \hat{\beta}_S) \leq q\, F(\alpha, q, n - p - 1) \qquad (14.6)$$

where q is the number of coefficients in β_S and $\hat{\beta}_S$ is the ols estimate of β_S. This specifies a q-dimensional ellipsoid. To get a 3D ellipsoid using the *R-code*, select the "Confidence Regions" item from the regression menu, and select three predictors. All the points in this plot will be on the surface of a 3D ellipsoid giving a confidence region. A view of this plot for the three variables *WT9*, *LG9*, and *ST9* is shown in Figure 14.4. Since the 3D plot is automatically centered and scaled, about all the user can see in this plot is orientation and elongation. When rotating this plot, one learns that the coefficient estimates for *WT9* and *LG9* are strongly and negatively correlated; the coefficient estimates for *LG9* and *ST9* are nearly uncorrelated, as are the coefficient estimates for *WT9* and *ST9*.

EXERCISES

14.1. Run the `demo-cr.lsp` demonstration as described in Section 14.2. Then, change the aspect ratio in the plot by making the window twice as long as it is high, and run the demonstration again. Describe any qualitative differences. Since the data have not changed, there are no *quantitative* differences, but perception of the plot can change.

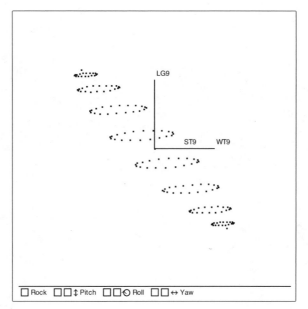

Figure 14.4. A 3D joint confidence region for the Berkeley Guidance Study for boys.

14.2. The file rat.lsp in the R-data folder contains the rat data described in Section 12.2.1.

14.2.1. Fit the regression as discussed in Section 12.2.1. Two of the predictors have relatively large *t*-values. What does this indicate? Get the correlation between these two estimated coefficients by selecting "Display Summaries" from the regression menu. Guess the shape of the joint confidence region for these two coefficients.

14.2.2. Draw the joint confidence region for the two coefficients studied in Exercise 14.2.1. Is the shape as you expected it to be?

14.2.3. Without removing the plot of the confidence region, draw a 3D added-variable plot for the two predictors identified. As you rotate this plot, examine it for any features that may help to understand the plot of the confidence region. Using the "O to e(O|H)" plot control may be helpful.

14.2.4. In the 3D added-variable plot, one point should have been identified as different from the others. How is it different? Select this point, and only this point, and then delete it by using the "Case Deletions" plot control. Not only is this plot updated, but the plot of the confidence region is updated as well. On the confidence region plot, the new ellipsoid and cross hairs are drawn in a different color, but the old ones are not removed. How does the confidence region change? What does this tell you about the case that was deleted?

APPENDIX A

The *R-code*

A.1 WHAT IS THE *R-code*?

The *R-code* is a computer program written in the *Xlisp-Stat* language. The computer code for *Xlisp-Stat* is required to run the *R-code*, and both are provided on the disks that come with this book. Users of the *R-code* can also use the language *Xlisp-Stat*.

The *R-code* is described in this book. The *Xlisp-Stat* language is a very powerful environment for statistical programming and is described in Tierney (1990). To use the *R-code*, you do not need to know how to program in *Xlisp-Stat* or any other language.

A.2 GETTING STARTED

This section describes how to get started using the *R-code*, depending on the type of computer you plan to use.

A.2.1 Macintosh Version

You will need at least 6 megabytes (mb) of memory (4 mb with System 6), and about 2 mb of free space on a hard disk. The *R-code* will not run acceptably on slower machines like the Mac Plus, Mac SE or PowerBook 100. Put the Macintosh disk in your floppy drive, double click on the icon called `Double click on me`, and follow the directions in the installer

225

program. Running the installer program will create a folder or directory called R-code Folder and, if you use System 7 or later, an item called Launch R-code will be added to your Apple Menu. You can choose in the installer to use a version of *Xlisp-Stat* that uses a math coprocessor.

To start the *R-code*, if you are using System 7 or later, select Launch R-code from the Apple menu. With any system, you can open the R-code Folder and double-click on the icon Xlisp-Stat 2.1 or Xlisp-Stat 2.1 881 if you are using the coprocessor version.

To remove the *R-code* from your disk, drag the R-Code Folder and the Launch R-Code file to the trash.

A.2.2 Windows Version

You will need Microsoft Windows 3.1 or newer, at least 4 mb of RAM, 2 mb of hard disk space, and VGA. Put the Windows disk into your disk drive. In the Windows Program Manager, select the item "Run..." from the "File" menu. In the text area of the resulting dialog, type a:\setup or b:\setup, as appropriate on your system, and then follow the directions in the setup program. This will create a directory with a name you specify, and put the program and data into it. The setup program will also create a group in the Program Manager called R-code, and an icon Launch R-code that will start the *Xlisp-Stat* program and load the *R-code*, and a second icon called Lspedit for a simple text editor.

No separate coprocessor version is available for Windows.

To remove the *R-code*, delete the R-Code directory from your disk and the group R-code from the Program Manager.

A.2.3 Unix Version

A separate Unix version is not distributed with this book. The *R-code* and *Xlisp-Stat* can be obtained electronically over the Internet by anonymous *ftp* to the machine *stat.umn.edu*, as follows:

```
% ftp stat.umn.edu
Connected to umnstat.stat.umn.edu.
Name (stat.umn.edu:): anonymous
Password: (type your email address)
ftp> cd pub/rcode
ftp> get rcode.readme.unix
ftp> quit
```

The file rcode.readme.unix gives further instructions.

If you already have *Xlisp-Stat* on your system, you can transfer the *R-code* from either the Macintosh or Windows version. You will need the files statinit.lsp, rcode.lsp, and the folders R-code and

`R-data`.   Put all these in the directory from which you plan to start *Xlisp-Stat*.

On most Unix systems, *Xlisp-Stat* is started by typing `xlispstat`.

A.2.4   Contents of the Disks

Both the Windows and Macintosh disks contain the following:

- A complete copy of the *R-code*, consisting of the files `statinit.lsp`, `rcode.lsp`, and the contents of the directory or folder `R-code`.

- A folder `R-data` containing all the data files and demonstration programs used in this book.

- A folder called `ALR` that includes all the data sets discussed in Weisberg (1985) in a format suitable for use with the *R-code*.

- All the files, examples, and data usually distributed with *Xlisp-Stat*. The files in the folders called `Data`, `Examples` are discussed in Tierney (1990). The folder `Glim` includes Tierney's code and documentation for fitting generalized linear models with *Xlisp-Stat*.

A.3   CUSTOMIZING THE *R-code*

When *Xlisp-Stat* starts, it loads a file called `statinit.lsp` in the directory or folder from which it is started. The `statinit.lsp` file included on the disks consists of the line `(load "rcode")` to load the *R-code*. If you already have a `statinit.lsp` file of your own, add this line to it.

Several constants are set in the file `rcode.lsp` and they can be changed by the user. For example, color rotating plots are not used with Macintoshes or PCs because color can significantly reduce performance. If you have a very fast processor or a fast graphics board, you might want to use color in rotating plots. Using an editor with the file `rcode.lsp`, change the value of `*use-color-in-3d-plots*` from `nil` to `t`. The file contains comments on other constants and how to change them. Remember to save the changed file without changing the name as a plain text file, not as a word processor file.

A.4   DATA FILES

A data file for use with the *R-code* can consist of data or both data and *Xlisp-Stat* commands, as illustrated in Section 1.4. Data can also be created in *Xlisp-Stat* and then a command can be typed to start the *R-code*. This is illustrated in Exercise 12.1.

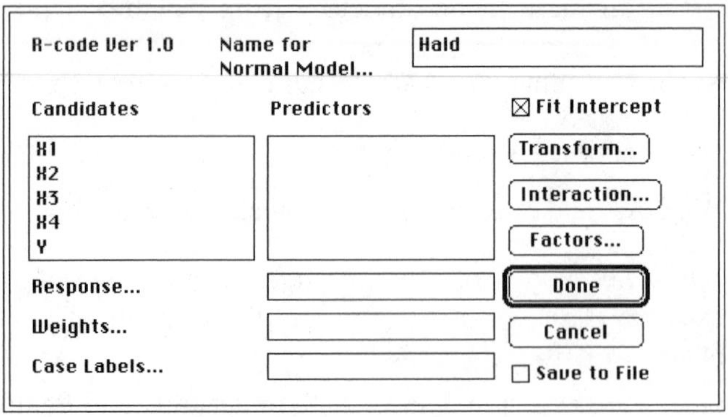

```
┌─────────────────────────────────────────────────────────────────┐
│  ┌───────────────────────────────────────────────────────────┐  │
│  │  R-code Uer 1.0    Name for        ┌──────────────────┐    │  │
│  │                    Normal Model... │ Hald             │    │  │
│  │                                    └──────────────────┘    │  │
│  │  Candidates           Predictors         ⊠ Fit Intercept   │  │
│  │  ┌───────────┐  ┌──────────────┐                           │  │
│  │  │ X1        │  │              │        ┌──────────────┐    │  │
│  │  │ X2        │  │              │        │ Transform... │    │  │
│  │  │ X3        │  │              │        └──────────────┘    │  │
│  │  │ X4        │  │              │        ┌──────────────┐    │  │
│  │  │ Y         │  │              │        │ Interaction..│    │  │
│  │  └───────────┘  └──────────────┘        └──────────────┘    │  │
│  │                                         ┌──────────────┐    │  │
│  │  Response...    ┌──────────────┐        │ Factors...   │    │  │
│  │                 └──────────────┘        └──────────────┘    │  │
│  │  Weights...     ┌──────────────┐        ┌──────────────┐    │  │
│  │                 └──────────────┘        │    Done      │    │  │
│  │  Case Labels... ┌──────────────┐        └──────────────┘    │  │
│  │                 └──────────────┘        ┌──────────────┐    │  │
│  │                                         │   Cancel     │    │  │
│  │                                         └──────────────┘    │  │
│  │                                         ☐ Save to File      │  │
│  └───────────────────────────────────────────────────────────┘  │
└─────────────────────────────────────────────────────────────────┘
```

Figure A.1. The standard regression dialog.

A.5   THE REGRESSION DIALOG

The regression dialog, like the one in Figure A.1, is used to set up a regression. Double clicking on a name in the "Candidates" list moves the name to the list of predictors. Clicking once on a name and then on an empty box for the response, weights, or case labels moves the name to that box. Double clicking on any of these boxes moves the name to the end of the list of candidates. The regression dialog has eight other items. At the top, you can type a name for the regression. This name will appear as the name of the regression menu. The program will always supply a default name. An intercept will be included unless "Fit Intercept" is unselected (the box is empty). Pushing any of the three buttons below the intercept control will give a dialog that can be used to modify data and then return to the regression dialog. The "Save to File" item, when checked, is used to create new data files. Push the "Done" button to create a model or the "Cancel" button to save any changes to the data but not create a model.

A.5.1   Transform Dialog

This dialog is used to create basic power transformations of existing variables. An example is shown in Figure A.2. If the "Done" button were now pushed, a new variable with values given by $(X_2 + 1)^{0.5}$ and a descriptive name would be created. If the "Log transform" button is pushed, then the value in the "Power" text area is ignored; a power of 0 is the same as the log transform. Press "Again" to transform another variable, "Done" to return to the regression dialog, or "Cancel" to return to the regression dialog without calculating a transformation. A more general transformation method is described in Section A.7.2.

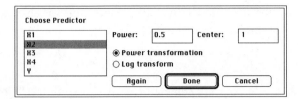

Figure A.2. The transformation dialog.

A.5.2   Interactions Dialog

This dialog is used to create interactions, which are elementary combinations of two predictors. It is selected by pushing the "Interactions..." button on the main regression dialog. An example of the interactions dialog is given in Figure 9.10. Select one item from each list in the dialog and an operation, either *, /, +, or −. A new variable will be created by applying the operation to the two variables selected. This is a slight generalization of the usual notion of interactions, where only the multiplication operation is used.

If either or both of the selected items are factors, then the only operation permitted is multiplication, *.

A.5.3   Factors Dialog

A *factor* is a collection of indicator variables that are used when a categorical predictor is to be included in a regression. Suppose, for example, that one predictor in a regression is "EyeColor." The variable in the data file could consist of the names of colors, such as "blue," "brown," "hazel," and "other," or it could consist of numbers to represent these categories. Since "EyeColor" has four unique values (levels), we could create up to four indicator variables. For example, one of the indicators would have the value 1 for every case with "blue" eyes and 0 for all other cases. In fitting a model with an intercept, the number of indicators required to fit a factor is 1 less than the number of levels.

In the "Factors..." dialog, double click on the names of all variables that you want to be factors to move them to the right list of selected variables. You can choose to have the factor consist of one indicator for each level or drop one level; either choice contains the same essential information. For example, you can choose "EyeColor" to be a factor with either three or four indicator variables. In the regression dialog you will see a name "EyeColor" and another name, "{F}EyeColor." "EyeColor" is the original, unchanged, variable, and "{F}EyeColor" is the factor created from "EyeColor." If you add "{F}EyeColor" to a model, then all the indicator variables in the factor are added to the model, and meaningful labels for them are created.

A.5.4 Save to File

If the "Save to File" item is selected, when you push the "Done" button you will get a dialog to specify a file name for saving the current set of variables and labels. The file can be read back into the *R-code* using the `load` command. The *R-code* will provide the `.lsp` suffix automatically.

A.5.5 Weighted Least Squares

To use weighted least squares, you must specify one of the variables to be case weights. If this variable is called w, then fitting will be done with the variance function σ^2/w. Each value of w must be positive. Weights are not used when superimposing fits or smooths, or removing linear trends from plots.

A.6 THE REGRESSION MENU

When a regression is specified in the *R-code*, a menu is added to the menu bar. The name of the menu and the name of the regression are the same. We call the menu the *regression menu*. Items in this menu that create graphs are discussed elsewhere in the book; look in the index under "Regression menu" for items not described here.

A.6.1 Display Fit

This produces the printed output obtained whenever you create a new regression. The first part of the output lists coefficient estimates, standard errors, and t-values. The second part lists several summary statistics including the estimated residual standard deviation. The last part of the output gives a summary analysis-of-variance table, with a line in the table for regression and a line for the residual. These are all standard calculations based on least squares or weighted least squares as discussed in any text on regression.

A.6.2 Display Summaries

This item prints univariate summary statistics for the predictors and the response and the sample correlation matrix. Also printed is the correlation matrix between the parameter estimates.

A.6.3 Sequential ANOVA

Selecting this item produces a sequential analysis-of-variance table. The predictors are added in the order they were chosen in the dialog.

A.6.4 Backward Elimination

Selecting this item fits a sequence of regressions obtained from the current regression by deleting at each step the current predictor with the smallest $|t\text{-value}|$, as long as this exceeds the cutoff chosen via a dialog.

A.6.5 New Model. . .

Selecting this item allows creation of a new regression using the standard regression dialog. The new regression will have its own menu and name.

A.6.6 Print. . .

Select this item to print data and derived case statistics in the text window. A dialog is used to select the quantities to print.

A.6.7 Remove Menu

Selecting this item removes the menu from the menu bar as well as any plots associated with the regression.

A.7 TYPING COMMANDS

Communication between the user and the *R-code* is done mostly by using the mouse and dialogs. There are some features of the *R-code*, and many features of *Xlisp-Stat*, that require typing.

The *R-code* creates the regression menu and a *regression object* that has the same name as the regression menu. For example, if the name of the menu is Hald, then the name of the regression object is also Hald. The name of the regression object is needed to extract information from the regression by sending the object *messages*, expressions that begin with a :, like :y. To print the values of the response variable, type

```
> (send hald :y)
(78.5 74.3 104.3 87.6 95.9 109.2 102.7 72.5 93.1 115.9 83.8
113.3 109.4)
```

To get the squared multiple correlation coefficient R^2, type

```
> (send hald :r-squared)
0.982376
```

This executes the code that computes R^2 for the model, and returns its value. If you enter the command

```
(/ (send hald :r-squared) (- 1 (send hald :r-squared)))
```

the value of $R^2/(1 - R^2)$ will be returned.

Table A.1 lists several useful messages that return values.

A.7.1   Getting Help

On-line help is available for most of the messages and functions. To get information about the :r-squared message, type

```
> (r-code :help :r-squared)
:R-SQUARED
Message args: ()
Returns the sample squared multiple correlation coefficient, R
squared, for the regression.
```

To get help for the iseq function, for example, type

```
> (help 'iseq)
ISEQ                                              [function-doc]
Args: (n &optional m)
With one argument returns a list of consecutive integers from 0
to N - 1. With two returns a list of consecutive integers from
N to M.  Examples: (iseq 4) returns (0 1 2 3)
                   (iseq 3 7)  returns (3 4 5 6 7)
                   (iseq 3 -3) returns (3 2 1 0 -1 -2 -3)
```

The command (apropos 'key) will print a list of all functions, messages, and objects that have the characters key in their names. For example, to get a list of all the cumulative distribution functions available in *Xlisp-Stat*, type

```
> (apropos 'cdf)
BINOMIAL-CDF
CHISQ-CDF
CAUCHY-CDF
GAMMA-CDF
BETA-CDF
NORMAL-CDF
BIVNORM-CDF
POISSON-CDF
F-CDF
T-CDF
```

Table A.1. Some Messages That Return Values

Message	Returns
:x	X, the $n \times p$ matrix of predictors.
:y	The response.
:residual-sum-of-squares	The residual sum of squares, RSS.
:num-cases	The number of cases.
:num-included	The number of included cases. An *included* case is used in regression calculations. A case usually becomes not included using the "Case Deletions" plot control.
:num-coefs	The number of predictors fit, including the intercept.
:df	The degrees of freedom for error.
:sigma-hat	The estimated residual standard deviation, $\hat{\sigma} = (RSS/df)^{1/2}$.
:r-squared	The squared multiple correlation R^2.
:coef-estimates	The coefficient estimates.
:xtxinv	$(X^T X)^{-1}$. To print a matrix in a readable form, you can use the command (print-matrix (send reg :xtxinv)), where reg is the name of the regression.
:coef-standard-errors	The standard errors of the coefficient estimates.
:leverages	Leverage values h_{ii}.
:fit-values	Fitted values $\hat{y}_i$.
:raw-residuals	$e_i = y_i - \hat{y}_i$.
:residuals	$w_i^{1/2}(y_i - \hat{y}_i)$ if weights are set and returns raw residuals otherwise.
:studentized-residuals	Studentized residuals, which are defined for included cases by $r_i = e_i / \hat{\sigma}(1 - h_{ii})^{1/2}$. For not included cases, $r_i = e_i / \hat{\sigma}(1 + h_{ii})^{1/2}$.
:ext-stud-res	Externally Studentized residuals, defined by $e_i / \hat{\sigma}_{(i)}(1 - h_{ii})^{1/2}$.
:cooks-distances	Cook's distances.
:local-influence	The direction of maximum curvature in the likelihood displacement for the coefficient vector when case weights are perturbed and the statistic $C_{\max}$ (Cook, 1986a).

To get help with the CDF of a *t*-distribution, type (help 't-cdf). To find the names of all functions based on the normal distribution, type (apropos 'norm) or (apropos 'normal).

A.7.2   General Transformations

The *R-code* allows limited transformation of variables using the transformation and interaction dialogs. The *Xlisp-Stat* language can be used to compute other transformations. Suppose we have defined a regression called reg with three variables named A, B, and C and we wanted to use the variable $\exp((A - B)/C)$. We can create a new variable,

```
> (def relvar (exp (/ (- a b) c)))
```

and then add it to the data available in the *R-code* using the :add-data message:

```
(send reg :add-data relvar "relvar")
```

This message requires two arguments, a list of numbers, and a name for the new variable. This variable is added to the list of candidates for reg and all its ancestors and descendants.

To add $\hat{y}^2$ to an existing model to perform Tukey's test of nonadditivity, first type

```
(send reg :add-data (^ (send reg :fit-values) 2) "Tukey")
```

and then use the "New Model..." item in the regression menu to create a model including *Tukey*. The *t*-value for this variable is the nonadditivity test.

As another example, suppose we have a variable *height* giving the height of corn plants and we want to convert this to categories so it could be used to define a factor. Suppose the categories of interest are *short*, less than 130 cm; *average*, between 130 cm and 160 cm; and *tall*, more than 160 cm. The typed command

```
(def cheight (cut height (list 130 160)
                  :values (list "short" "average" "tall")))
```

will create a new predictor *cheight* that consists of the values "short" for *height* $\leq$ 130, "average" for 130 < *height* $\leq$ 160, and "tall" for *height*

> 160. This can be added to the regression problem using the `:add-data` message, converted to a factor using the "Factors..." dialog, and then used in a regression model.

This last transformation made use of the `cut` function. The documentation for this function is

```
> (help 'cut)
CUT                                                    [function-doc]
Message args: x cutpoints &key (values (iseq 0 (length cut-
points))) Returns a list of the same length as x by discretiz-
ing x into 1+(length cutpoints) categories.  Cutpoints must
be ordered from smallest to largest. The keyword :values, if
present, gives the labels for the categories.
```

A.7.3 Renaming or Deleting Variables

Variables can be renamed or deleted. Assuming you have a regression object named `reg`, type the command

```
(send reg :change-data)
```

In the resulting dialog, select the old name from the list, and either type the new name or push the "Delete from dataset" button. The new names will appear immediately in the "Plot of..." and "New Model..." dialogs, but they will not appear in any existing model. The new names will be saved if you use the "Save to File" option in the regression menu.

A.7.4 Linear Combinations

The `:lin-combination` message is used to compute an estimate of a linear combination of parameters and its standard error. This message requires $p+1$ arguments $x_0, x_1, \ldots, x_p$ for a linear model with p predictors and an intercept. The function returns a list of two elements, the linear combination $\hat{\beta}_0 x_0 + \hat{\beta}_1 x_1 + \cdots + \hat{\beta}_p x_p$ and the standard error of this linear combination of the coefficient estimates. For example, in the Hald data in Table 1.2 of Chapter 1, type

```
> (send hald :lin-combination 1 20 20 20 40)
(99.9065 4.90003)
```

The fitted value at $x_1 = x_2 = x_3 = 20$ and $x_4 = 40$ is 99.9065, and its standard error is about 4.9.

As a second example, suppose the contrast $\beta_1 - \beta_2$ were of interest. One could compute

```
> (send hald :lin-combination 0 1 -1 0 0)
(1.04094 0.230733)
```

so the estimate is 1.04 with standard error 0.23. If the intercept is not included in the model, then a multiplier for it should not be supplied to this message.

A.7.5   The Plot Message

The :plot message is used to draw plots of quantities that do not appear in the "Plot of..." list. Suppose we have two regression models called reg and reg1. To plot the residuals from the fit in reg on the horizontal axis against the fitted values from the fit in reg1 on the vertical axis, type

```
(send reg :plot :residuals (send reg1 :fit-values))
```

The :plot message requires at least one argument. Each argument can be either a list of n numbers, such as (send reg1 :fit-values), or the name of one of the messages in Table A.1 or any other message that returns a list of n numbers, where n is the number of cases. Specifying one argument produces a histogram, two a scatterplot, three a rotating plot, and more than three a scatterplot matrix. All arguments that are names of messages will be updated when the data are changed by deleting a case. In the last example a message was used to compute one axis, but the command (send reg1 :fit-values) produced a list of numbers, so the horizontal axis will be updated but the vertical axis will not be updated.

Suppose you wanted a plot of $\{(1 - h)\hat{y}, |e|^{1/2}\}$, a plot for nonconstant variance. This can be done in two ways. Typing

```
(send reg :plot
        (* (- 1 (send reg :leverages)) (send reg :fit-values))
        (sqrt (abs (send reg :residuals))))
```

will work, but the plot will not be updated if cases are deleted. This can be overcome by defining messages that compute these values, as follows:

```
(defmeth regression-model-proto :lev-hat ()
  (* (- 1 (send self :leverages)) (send self :fit-values)))
(defmeth regression-model-proto :sqrtres ()
  (sqrt (abs (send self :residuals))))
(send reg :plot :lev-hat :sqrtres)
```

and this plot will be updated.

Suppose you wanted to use the :lev-hat message regularly. Here are two steps that can make Lev-hat appear in the list of candidates for plots. First, append the code for this message to the file updates.lsp in the R-code folder. This message will then be defined every time you start the *R-code*. Then, in the file rcode.lsp, modify the constant *default-plotof-list* to include the value :lev-hat. That's all there is to it.

A.7.6   Naming Plots Produced by the *R-code*

Every plot created by the *R-code* is an object. If you know the plot's name, you can send it messages to add lines or curves, add text, change labels and tick marks, and so on. The messages for these changes are described in Tierney (1990). If you have a regression object called hald, then the message

```
(def p (send hald :last-graph))
```

will assign the name p to the last graph created using the hald menu. You could add a line to p, with intercept 1 and slope 3, by sending the message

```
(send p :abline 1 3)
```

A.8   ERROR MESSAGES

The *R-code* occasionally produces error messages in dialogs and in the text window. Error messages in dialogs are for specific problems and are self-explanatory. Error messages in the text window are for more generic problems, like dividing by zero or attempting to use a nonnumeric quantity in arithmetic. This section provides a brief listing of some error messages.

error: misplaced right paren A typed expression has an extra right parenthesis.

error: can't assign to a constant *Xlisp-Stat* uses a few names like pi, e, and t. You cannot use these names for your own variables.

error: not a number - ABC The characters ABC were used in a place where the program was expecting a number.

error: unbound variable - ABC The variable ABC is not defined.

`error: sequences of different lengths` Lists of incompatible lengths were specified in some operation. This can occur if you try to plot a list of five elements against a list of six elements or if the number of cases for the predictor differs from the number for the response.

`error: illegal zero argument` *Xlisp-Stat* detected an attempt to divide by zero.

`error: too few arguments` or `error: too many arguments` Function or message was called with the wrong number of arguments.

`error: not a list - NIL` A function or message expected a list of values as an argument, but the value it was passed is `NIL`.

`error: dimensions do not match` You have attempted to do arithmetic with lists or matrices of incompatible lengths. This will occur in the (`rcode`) function if the number of cases for the predictor does not equal the number for the response. It will also occur in matrix multiplication if the matrices are of the wrong sizes.

`error: arguments not all the same length` An argument to a function or message is of the wrong length.

`error: not a valid graph address - try reallocating the object` A message was sent to a graph that no longer exists. If you get this message while trying to draw a graph using a menu item in the *R-code*, use the "New Model..." item to create a new copy of the regression and draw the plot from the new menu.

`error: not enough memory to allocated to color port` This probably indicates that your computer has insufficient memory to use color in plots. You can try setting the constant `*use-color-in-3d-plots*` to `nil` in the file `rcode.lsp`.

`error: insufficient node space` The program has insufficient memory. Closing plots may help. On the Macintosh, you can change the memory allocation using the "Get Info" item about *Xlisp-Stat* in the finder.

APPENDIX B

Copyrights

Copyright law protects the rights of the creators of intellectual material. The copyright for this book is held by the publisher, John Wiley & Sons, Inc. The rights retained by the publisher are outlined on the copyright page.

Like a book, computer programs are also protected by copyright. The *R-code* is written in the *Xlisp-Stat* language, defined by Tierney (1990). *Xlisp-Stat* in turn is based on *Xlisp*. The copyright for *Xlisp-Stat* is held by its author, Luke Tierney, while the copyright for *Xlisp* is held by its author, David Betz. While retaining the copyright, these authors have granted permission to "copy, modify, distribute, and sell this software and its documentation for any purpose." The complete license statement is given in the file COPYING on both the disks included with this book.

The copyright to the *R-code* is held by its authors, R. Dennis Cook and Sanford Weisberg. The *R-code* consists of all the files distributed in the directories R-code and R-data that include a copyright notice. License to use the *R-code* is granted as follows:

- If the purchaser of this book is an individual, then that person may use Version 1 of the *R-code* on any computer system.

- If the purchaser of this book is a library, then any person who checks out this book may use Version 1 of the *R-code* while in possession of the book.

239

- If the purchaser of this book is any other entity, then one user at a time affiliated with that entity may use Version 1 of the *R-code*. For several simultaneous users (e.g., in a computer laboratory), multiple copies of this book must be purchased.

In particular, this license does not grant permission to give copies of the *R-code* to anyone. If you sell this book, then you also sell your right to use the *R-code*.

Except when otherwise stated in writing, the copyright holders and/or other parties provide the program "as is" without warranty of any kind, either expressed or implied, including, but not limited to, the implied warranties of merchantability and fitness for a particular purpose. The entire risk as to the quality and performance of the program is with you. Should the program prove defective, you assume the cost of all necessary servicing, repair, or correction.

References

Altman, N. S. (1992). An introduction to kernel and nearest-neighbor nonparametric regression. *American Statistician*, 46, 175–185.

Atkinson, A. C. (1981). Robustness, transformations and two graphical displays for outlying and influential observations in regression. *Biometrika*, 68, 13–20.

Atkinson, A. C. (1985). *Plots, Transformations and Regression*. Oxford: Oxford University Press.

Becker, R. A., and Cleveland, W. (1987). Brushing scatterplots. *Technometrics*, 29, 127–142, reprinted in Cleveland, W., and McGill, M. (1988), *op. cit.*, 201–224.

Box, G. E. P., and Cox, D. R. (1964). An analysis of transformations. *Journal of the Royal Statistical Society, Series B*, 26, 211–246.

Breiman, L., and Friedman, J. (1985). Estimating optimal transformations for multiple regression and correlation. *Journal of the American Statistical Association*, 80, 580–597.

Breusch, T. S., and Pagan, A. R. (1979). A simple test for heteroscedasticity and random coefficient variation. *Econometrica*, 47, 1287–1294.

Brillinger, D. (1983). A generalized linear model with "Gaussian" regression variables. In Bickel, P. J., Doksum, K. A., and Hodges Jr., J. L., eds., *A Festschrift for Erich L. Lehmann*. New York: Chapman & Hall, 97–114.

Brinkman, N. D. (1981). Ethanol fuel—A single-cylinder engine study of efficiency and exhaust emissions. *SAE Transactions*, 90, No. 810345, 1410–1424.

Carroll, R. J., and Ruppert, D. (1988). *Transformations and Weighting in Regression*. New York: Chapman & Hall.

Chambers, J., Cleveland, W., Kleiner, B., and Tukey, P. (1983). *Graphical Methods for Data Analysis*. New York: Chapman & Hall.

Chen, C. F. (1983). Score tests for regression models. *Journal of the American Statistical Association*, 78, 158–161.

Cleveland, W. (1979). Robust locally weighted regression and smoothing scatterplots. *Journal of the American Statistical Association*, 74, 829–836.

Cleveland, W., and McGill, M. (1988). *Dynamic Graphics for Statistics*. New York: Chapman & Hall.

Cochran, W. G., and Cox, G. (1957). *Experimental Designs*, 2nd ed. New York: Wiley.

Cook, R. D. (1977). Detection of influential observations in linear regression. *Technometrics*, 19, 15–18.

Cook, R. D. (1986a). Assessment of local influence (with discussion). *Journal of the Royal Statistical Society, Series B*, 48, 134–169.

Cook, R. D. (1986b). Discussion of "Influential observations, high leverage points and outliers in linear regression," by S. Chatterjee and A. Hadi, *Statistical Science*, 1, 379–416.

Cook, R. D. (1993). Exploring partial residual plots. *Technometrics*, 35, 351–362.

Cook, R. D. (1994). On the interpretation of regression plots. *Journal of the American Statistical Association*, 89, 177–189.

Cook, R. D., Hawkins, D., and Weisberg, S. (1992). Comparison of model misspecification diagnostics using residuals from least mean of squares and least median of squares fits. *Journal of the American Statistical Association*, 87, 419–424.

Cook, R. D., and Nachtsheim, C. (1994). Re-weighting to achieve elliptically contoured covariates in regression. *Journal of the American Statistical Association*, 89, June.

Cook, R. D., Peña, D., and Weisberg, S. (1988). The likelihood displacement: A unifying principle for influence measures. *Communications in Statistics, Part A—Theory and Methods*, 17, 623–640.

Cook, R. D., and Weisberg, S. (1982). *Residuals and Influence in Regression*. London: Chapman & Hall.

Cook, R. D., and Weisberg, S. (1983). Diagnostics for heteroscedasticity in regression. *Biometrika*, 70, 1–10.

Cook, R. D., and Weisberg, S. (1989). Regression diagnostics with dynamic graphics (with discussion). *Technometrics*, 31, 277–311.

Cook, R. D., and Weisberg, S. (1990a). Three dimensional residual plots. In Berk, K., and Malone, L., eds., *Proceedings of the 21st Symposium on the Interface: Computing Science and Statistics*. Washington: American Statistical Association, 162–166.

Cook, R. D., and Weisberg, S. (1990b). Confidence curves for nonlinear regression. *Journal of the American Statistical Association*, 85, 544–551.

Cook, R. D., and Weisberg, S. (1991). Comment on Li (1991). *Journal of the American Statistical Association*, 86, 328–332.

Cook, R. D., and Weisberg, S. (1994). *ARES* plots for generalized linear models. *Computational Statistics and Data Analysis*. 17, 303–315.

Cook, R. D., and Weisberg, S. (submitted). Transforming a response variable for linearity.

Cook, R. D., and Wetzel, N. (1993). Exploring regression structure with graphics (with discussion). *TEST* 2, 1–57.

Cox, D. R. (1958). *The Planning of Experiments*. New York: Wiley.

Doll, R. (1955). Etiology of lung cancer. *Advances in Cancer Research*, Vol. 3, reprinted in Report of the Advisory Committee to the Surgeon General (1964), *Smoking and Health*. Washington, DC: U.S. Government Printing Office, 164.

Duan, N., and Li, K. C. (1991). Slicing regression: A link-free regression method. *Annals of Statistics*, 19, 505–530.

Eaton, M. L. (1986). A characterization of spherical distributions. *Journal of Multivariate Analysis*, 20, 272–276.

Enz, R. (1991). *Prices and Earnings around the Globe*. Zurich: Union Bank of Switzerland.

Ezekiel, M. (1924). A method of handling curvilinear correlation for any number of variables. *Journal of the American Statistical Association*, 19, 431–453.

Ezekiel, M. (1941). *Methods of Correlation Analysis*, 2nd ed. New York: Wiley.

Federer, W. T. (1955). *Experimental Designs*. New York: MacMillian.

Federer, W. T., and Schlottfeldt, C. S. (1954). The use of covariance to control gradients in experiments. *Biometrics*, 10, 282–290.

Fox, J. (1991). *Regression Diagnostics*. Newberry Park, CA: Sage.

Fisherkeller, M. A., Friedman, J. H., and Tukey, J. W. (1974). PRIM-9: An interactive multidimensional data display and analysis system, reprinted in Cleveland, W., and McGill, M. (1988), *op. cit.*, 91–110.

Freund, R. J. (1979). Multicollinearity etc., some "new" examples. *Proceedings of ASA Statistical Computing Section*. Washington, DC: American Statistical Association, 111–112.

Gnanadesikan, R. (1977). *Methods for Statistical Analysis of Multivariate Data*. New York: Wiley.

Hall, P., and Li, K. C. (1993). On almost linearity of low dimensional projections from high dimensional data. *Annals of Statistics*, 21, 867–889.

Härdle, W. (1990). *Applied Nonparametric Regression*. Cambridge: Cambridge University Press.

Hastie, T., and Tibshirani, R. (1990). *Generalized Additive Models*. New York: Chapman & Hall.

Hernandez, F., and Johnson, R. A. (1980). The large sample behavior of transformations to normality. *Journal of the American Statistical Association*, 75, 855–861.

Hinkley, D. (1975). On power transformations to symmetry. *Biometrika*, 62, 101–111.

Hinkley, D. (1985). Transformation diagnostics for linear models. *Biometrika*, 72, 487–496.

Hoaglin, D., and Welsch, R. (1980). The hat matrix in regression and ANOVA. *American Statistician*, 32, 17–22.

Huber, P. (1981). *Robust Statistics*. New York: Wiley.

John, J. A., and Draper, N. R. (1980). An alternative family of power transformations. *Applied Statistics*, 29, 190–197.

Li, K. C. (1991). Sliced inverse regression for dimension reduction (with discussion). *Journal of the American Statistical Association*, 86, 316–342.

Li, K. C. (1992). On principal Hessian directions for data visualization and dimension reduction: Another application of Stein's lemma. *Journal of the American Statistical Association*, 87, 1025–1039.

Li, K. C., and Duan, N. (1989). Regression analysis under link violation. *Annals of Statistics*, 17, 1009–1052.

McCulloch, R. (1993). Fitting regression models with unknown transformations using dynamic graphics. *The Statistician*, 42, 153–160.

Mickey, R., Dunn, O. J., and Clark, V. (1967). Note on the use of stepwise regression in detecting outliers. *Computers and Biomedical Research*, 1, 105–109.

Mosteller, F., and Tukey, J. W. (1977). *Data Analysis and Regression*. Reading: Addison-Wesley.

Schott, J. R. (1994). Determining the dimensionality of sliced inverse regression. *Journal of the American Statistical Association*, 89, 141–148.

Scott, D. (1992). *Multivariate Density Estimation*. New York: Wiley.

Searle, S. R. (1988). Parallel lines in residual plots. *American Statistician*, 43, 211.

Stuetzle, W. (1987). Plot windows. *Journal of the American Statistical Association*, 82, 466–75. Reprinted in Cleveland, W., and McGill, M. (1988), *op. cit.*, 225–246.

Tierney, L. (1990). *Lisp-Stat: An Object-oriented Environment for Statistical Computing and Dynamic Graphics*. New York: Wiley.

Tuddenham, R. D., and Snyder, M. M. (1954). Physical growth of California boys and girls from birth to age 18. *California Publications on Child Development*, 1, 183–364.

Tufte, E. (1974). *Data Analysis for Politics and Policy*. Englewood Cliffs, NJ: Prentice-Hall.

Velleman, P. (1982). Applied nonlinear smoothing. In Leinhardt, S., ed, *Sociological Methodology 1982*. San Francisco: Jossey-Bass, 141–177.

Verbyla, A. P. (1993). Modelling variance heterogeneity: Residual maximum likelihood and diagnostics. *Journal of the Royal Statistical Society, Series B*, 55, 493–508.

Weisberg, S. (1985). *Applied Linear Regression*, 2nd ed. New York: Wiley.

Wood, F. S. (1973). The use of individual effects and residuals in fitting equations to data. *Technometrics*, 15, 677–695.

Index

WILEY SERIES IN PROBABILITY
AND MATHEMATICAL STATISTICS

ESTABLISHED BY WALTER A. SHEWHART AND SAMUEL S. WILKS

Editors

Vic Barnett, Ralph A. Bradley, Nicholas I. Fisher, J. Stuart Hunter, J. B. Kadane, David G. Kendall, Adrian F. M. Smith, Stephen M. Stigler, Jozef L. Teugels, Geoffrey S. Watson

*Now available in a lower priced paperback edition in the Wiley Classics Library.

*Now available in a lower priced paperback edition in the Wiley Classics Library.

*Now available in a lower priced paperback edition in the Wiley Classics Library.